Bird of Paradise Screen—HT-1, HT-11, HT-12—With Trumpet Vine lamp. Miller residence

Interior door using bevels and brass came.   LoGrasso residence

Beveled entry sidelights with brass came.   Anderson residence

Two story entry windows designed with privacy and sun control in mind.
Tawfik residence

Living room cathedral windows. French residence

Entry windows with Trumpet vine and trellis. Dr. Brownell residence

Dining room windows (sliding) with design concentrated in lower half to soften view of rooftops. Mayo residence

Entry windows. Wong residence

Kitchen windows in peach and green bevels with clear double glue chip background. Ikeda residence

Conference room window depicting company's interest. Elektra Power Corporation

Beveled entry window. Ikeda residence

Entry skylight, 8′ diameter, of magnolia blossoms and hummingbirds. Magleby residence

Entry sidelights—HT-8—Dogwood and Butterflies. Arnold residence

Peach and green beveled cabinet doors. Ikeda residence

Cabinet doors updated with 2″x2″ bevels on the diagonal and double glue chip. 4′x8′ skylight (in five sections) with 2″ bevels, double glue chip and borders to coordinate with tile and hardwood floor. Miller residence

Kitchen cabinet doors with custom 2″ bevels in bright colors against a clear antique seedy background. Brass came used to enhance existing brass fittings in kitchen. Moore residence

Dining room with six-panel folding doors of clear antique and bevels. Doors are energy savers in the winter and non-restrictive dividers throughout the year. Cabinet doors are also clear antique. Lamp is Louis XIV. Miller residence.

Living room 4 panel screen featuring yellow bevels surrounding design of peonies and willows.   Kauffman residence

Pheasant and hunting dogs make a beautiful statement in the family room.   Bullock residence

Game table.   Miller residence

Family room fireplace screen with Magnolia lamp. Miller residence

Three panel Peacock screen for garden room. Miller residence

Water Irises—HT-5—Designed as lightbox for end of long hallway. Small Christmas tree lights provide illumination. Kauffman residence

Wild Rose—HT-7—window designed with unequal border to camouflage solar pipes beyond. Miller residence

Stained glass for the bath. Anderson residence

Wisteria vines for the bath. Jackson residence

Bath window with sections of the wallpaper, sink and waste basket enlarged, rearranged and completed in stained glass. Moore residence

Bedroom windows with beveled fleur de lis and opaque glass for privacy. Fabric art warm window coverings. White Peony lamp. Miller residence

Magnolia blossoms adorn the bedroom. Jackson residence

Fabric Art—Place mat utilizing section of "Spring". Bradley residence
—Wall hanging of section of "Winter". Bradley residence
—Hooked rug and stained glass carousel horses are worked from HT-9—Carousel Horse. Jackson residence and Moss residence
—Water Iris—HT-5—in needlepoint. Dickman residence
—Caftan with morning glories (page 86). Mulvey residence

Carousel horse in fabric art and stained glass. Jewels, gold and silver threads used in fabric art carousel horse. Faceted amethyst, topaz, garnet and peridot set in stained glass carousel horse.

Wisteria stained glass mirror and wearable art (page 89).
Barnett residence
model—Paula Elmore

Iris stained glass mirror designed to compliment bedding and drapes. French residence

Carousel Horse. Jackson residence

Three Irises—HT-6—Beveled border enhances this delightful window. Blumenthal residence

HOUSE TOURS
by
Judy Miller

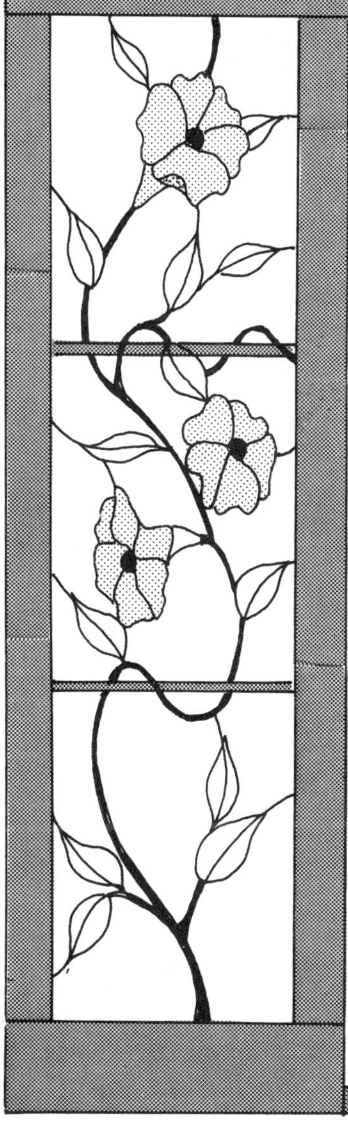

HOUSE TOURS is designed as a room by room guide for stained glass and needle arts in the home.

Stained glass windows and needle arts are wonderful additions to all homes. the designs are equally suited for cabinet doors, table tops, room dividers, screens, wall hung "pictures", mirrors, fire screens, free form three-dimensional pieces and lamps of all styles.

The designs recreated in threads make beautiful pillows, bedspreads, window coverings, hooked rugs, chair coverings, room dividers, screens and great wearable art.

Please join in this HOUSE TOUR of ideas.

A VERY SPECIAL THANK YOU TO THE FOLLOWING ARTISTS FOR HELPING TO MAKE THIS BOOK POSSIBLE

STAINED GLASS ARTISTS

Al Bullock
Bruce Finnie
Jim Forsell
Jane French
Mary Jackson
Marcelle Kauffman
Deanne Klages
Gene Mayo
Ome Stark

STITCHERY ARTISTS

Jean Barnett
Nancy Becket
Cyndy Bradley
Carol Dickman
Jane and Ted Moss
Trish Mulvey

HOMES PHOTOGRAPHED

Mr. and Mrs. Joseph Anderson
Mr. and Mrs. Gordon Arnold
Mr. Robert Blumenthal
Dr. and Mrs. Robert Brownell
Mr. Al Bullock
Mr. and Mrs. Dixon Collins
Elektra Power Corporation
Ms. Jane French
Ms. Hedy Herr
Mr. and Mrs. Ikira Ikeda
Mr. and Mrs. Al Jackson
Mr. and Mrs. Daniel Kauffman
Mr. and Mrs. Vincenzo Lo Grasso
Dr. and Mrs. Kay Magleby
Mr. and Mrs. Gene Mayo
Ms. Judy Miller
Mr. and Mrs. William Moore
Mr. and Mrs. Wageeh Tawfik
Mr. and Mrs. Al Wong

Photography Warren Hukill
Color Kedie/Orent, Inc.
Screens Wooden Visions, Inc.
Judy Miller
Firescreens & wrought iron furniture Tony Archuleta
Custom Bevels Ted Moss
Model Paula Elmore

—And to Bruce Finnie, who not only generously provided expert advice and help in the layout and printing of this book, but provided many of the beautiful lamps as well.

First Edition—Published in the United States by Judy Miller Publications
2979 Alexis Drive, Palo Alto, California 94304

ISBN: 0-912833-02-5

Distribution—JUDY MILLER PUBLICATIONS,
2979 ALEXIS DRIVE,
PALO ALTO, CALIFORNIA 94304
(415) 948-7349

THE HOUSE TOURS PATTERNS

HT-1 Spring HT-2 Summer HT-3 Autumn HT-4 Winter

HT-5 Water Iris

HT-6 Three Iris

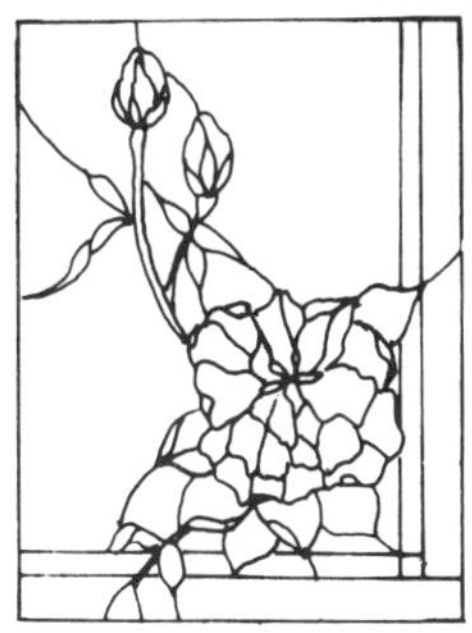

HT-7 Wild Roses

HT-8 Dogwood & Butterflies

HT-9 Carousel Horse

HT-10
Bird of Paradise "A"

HT-11
Bird of Paradise "B"

HT-12
Bird of Paradise "C"

HT-1 SPRING 12″x48″
HT-2 SUMMER 12″x48″
HT-3 AUTUMN 12″x48″
HT-4 WINTER 12″x48″
HT-5 WATER IRIS 11½″x37″
HT-6 THREE IRIS 25½″x34½″
HT-7 WILD ROSES 22′x32½″
HT-8 DOGWOOD & BUTTERFLIES 7½″x32½″ (each)
HT-9 CAROUSEL HORSE 30″ Diameter
HT-10 BIRD OF PARADISE "A" 18″x28″
HT-11 BIRD OF PARADISE "B" 18″x28″
HT-12 BIRD OF PARADISE "C" 18″x28″

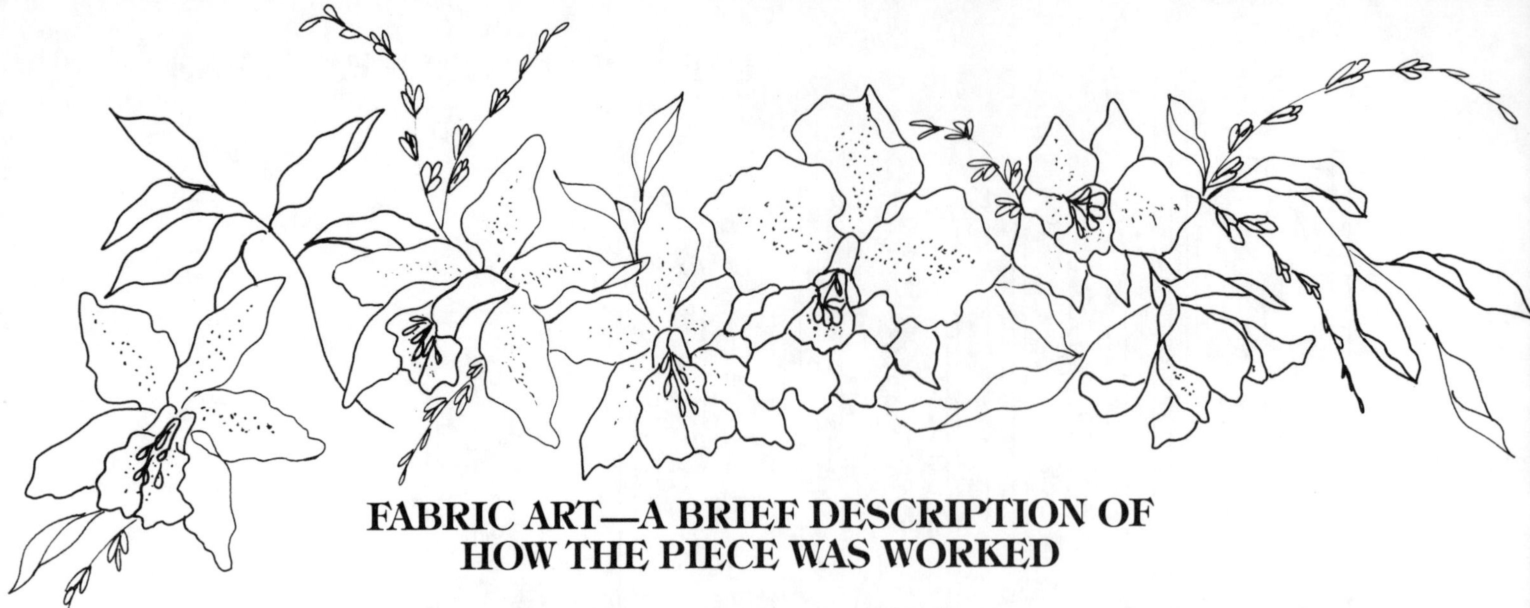

FABRIC ART—A BRIEF DESCRIPTION OF HOW THE PIECE WAS WORKED

Jacket — "Wisteria" (adapted from page 89) Stichery worked in two layers. Bottom layer is four shades of perle cotton, laid horizontally, catching the top layer of wool weave. Top layer is single and double layer floss in five shades, stitched to enhance the sculpted shape of the wisteria. Vine is double strands of #5 DMC, couched with floss of varying shades to create depth. Trellis is #3 cotton perle satin stitched over a #5 outline. Designed and stitched by Jean Barnett.

Wall Hanging — "Carousel Horse" (HT-9) Stitched on black silk. Beads, jewels and braid beautifully used on tack, saddle and blanket. Silver threads outline mane and tail. Designed and stitched by Jean Barnett.

Place Mat — "Spring" (HT-1) Assissi counted work on Jobelan. Adapted and stitched by Cyndy Bradley.

Wall Hanging — "Winter" (HT-4) Sampler of threads and techniques on Novara. Adapted and stitched by Cyndy Bradley.

Wall Hanging — "Water Iris" (HT-5) Needlepoint study in blue. Adapted and stitched by Carol Dickman.

Rug — "Carousel Horse" (HT-9) Rug worked by Jane and Ted Moss.

Caftan — "Morning Glories" (page 86) Machine applique on linen. Adapted and stitched by Trish Mulvey.

ADJUSTING A DESIGN FOR YOUR PARTICULAR REQUIREMENTS

Almost any design can be modified to fit an existing opening. For example, if the desired design is too small for the opening, consider using the design as a circular or oval medallion and design a field and border around this medallion to the dimensions required.

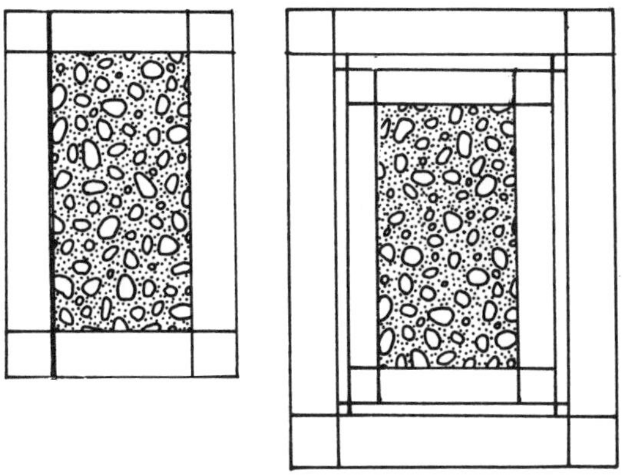

Enlarging with equal borders

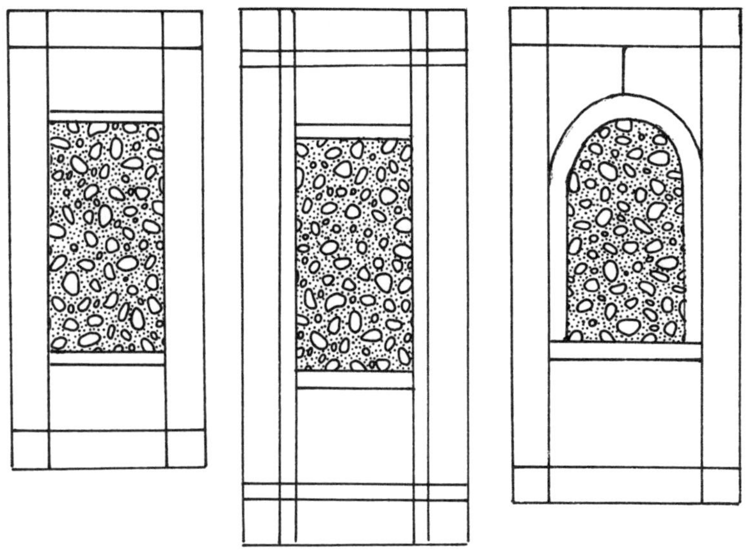

Enlarging with unequal borders

—or use a repeat of one design or a series of three for long sidelights or transoms.

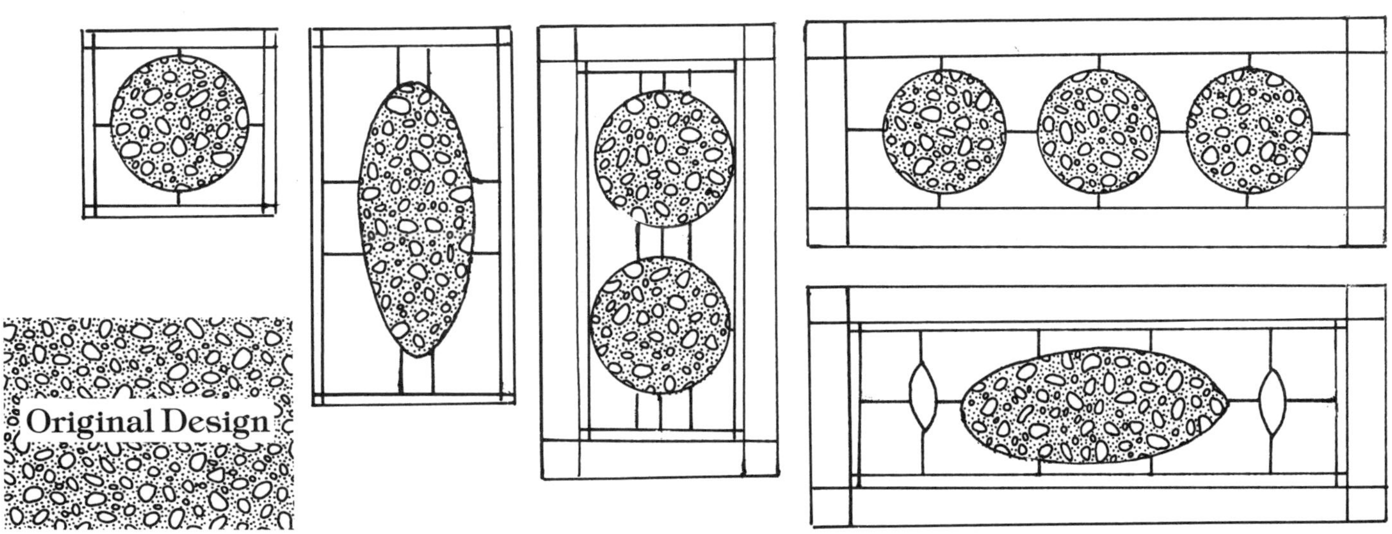

Flower

ENTRY WINDOWS are usually door panels, transoms or one or two sidelights.

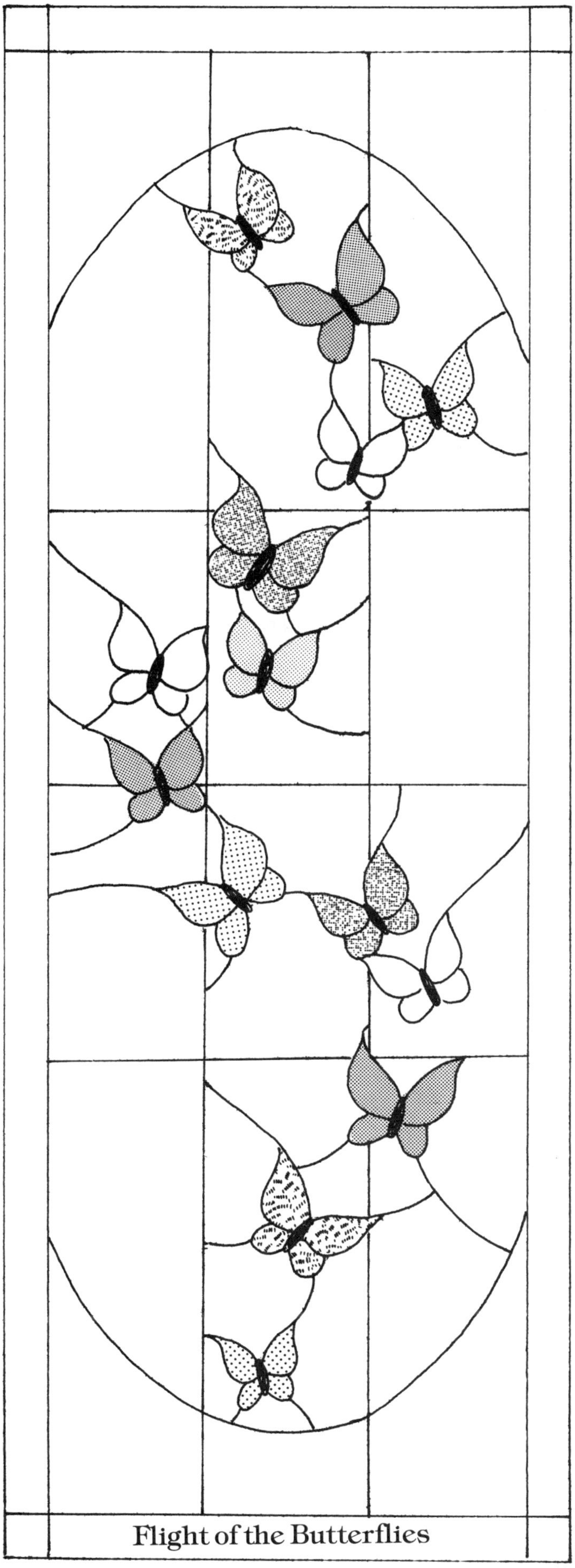

Flight of the Butterflies

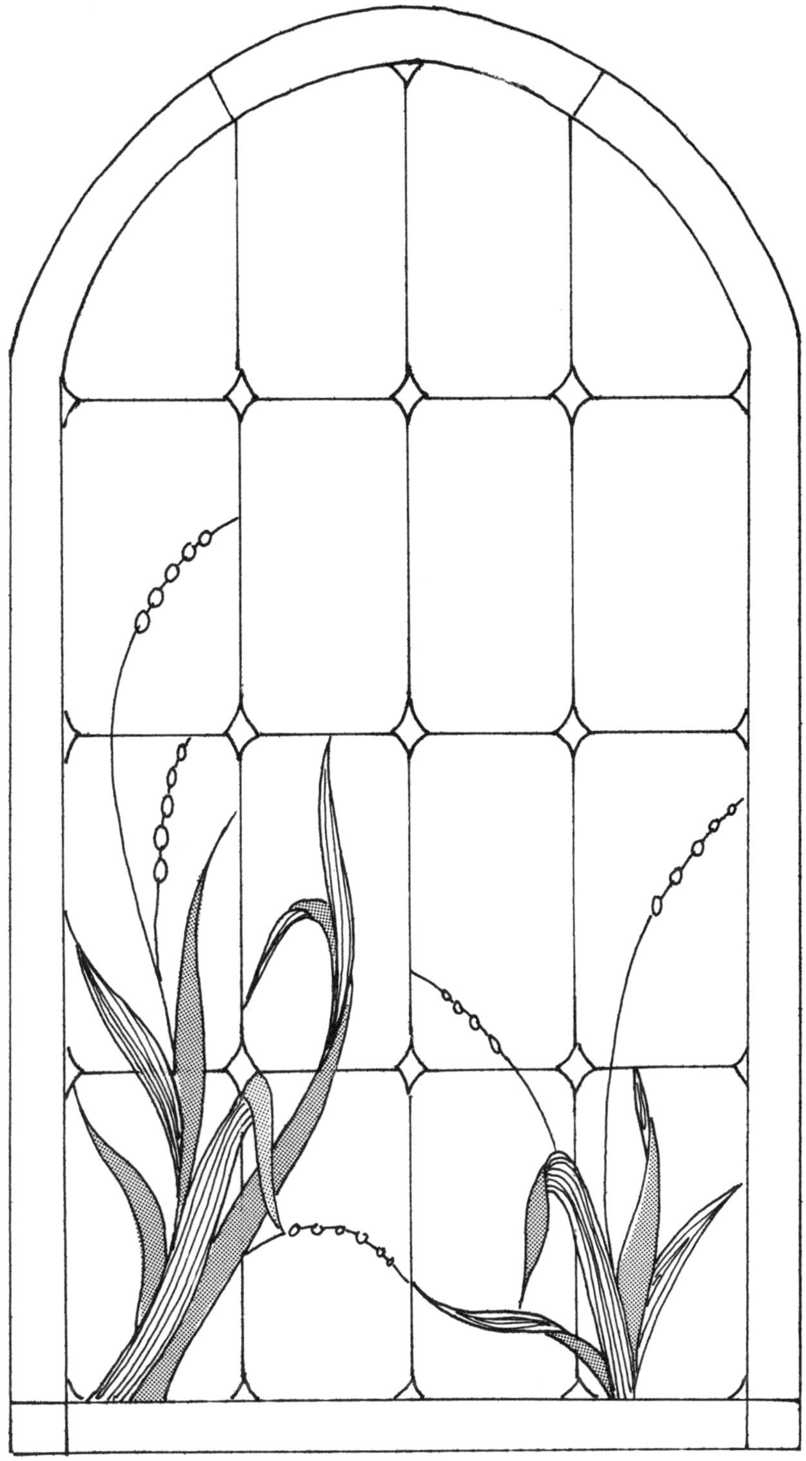

Beveled diamonds and jewel flowers enhance this design

Bamboo sidelights

Bird of Paradise sidelights

Jewels add drama to this design.

Sidelights easily enlarged with addition of borders

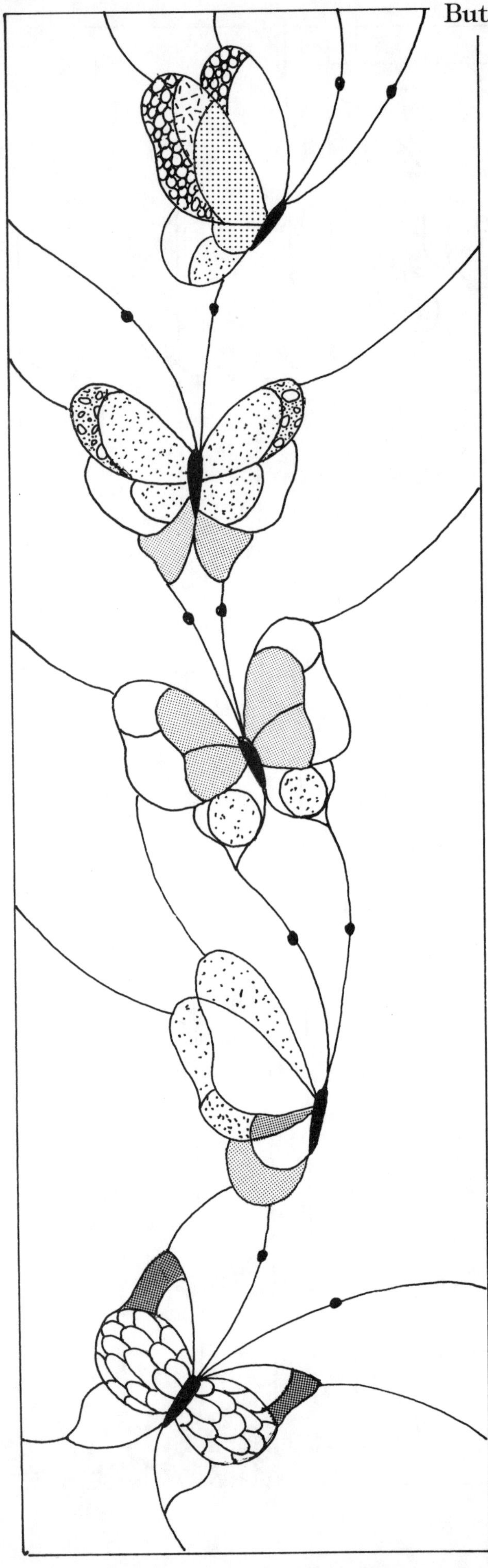

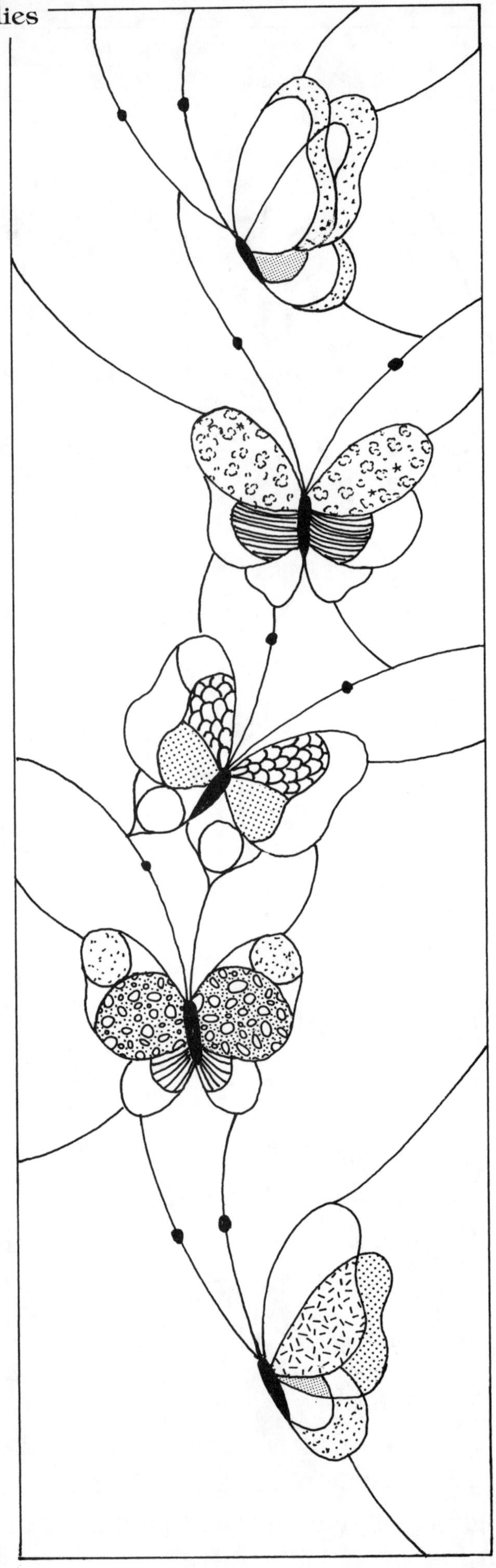

Apple Blossom Time

LIVING ROOM WINDOWS are a wide variety of sizes and shapes. Following are design ideas for unusual window shapes.

Stained glass hung as "paintings" on the wall add a great dramatic touch to the living room. Properly selected opalescent glass is recommended. If you can build a lightbox for the piece, this is even more dramatic.

Consider adding stained glass tabletops to your living room decor. They are beautiful and have the wonderful advantage of being so practical. A ¼-inch plate glass should be placed on top of the stained glass to protect it as well as to provide an even surface for items placed on the table. A good support system of zinc bracing, either worked into the design or soldered onto the backside upon completion will prevent the tabletop from bowing. If space allows, sandwiching the stained glass between two pieces of plate glass is another approach. This has the advantage of protecting the underside from possible damage.

If you have interesting prints or a series of prints, why not frame them in stained glass rather than the usual wood. The use of brass came is most effective.

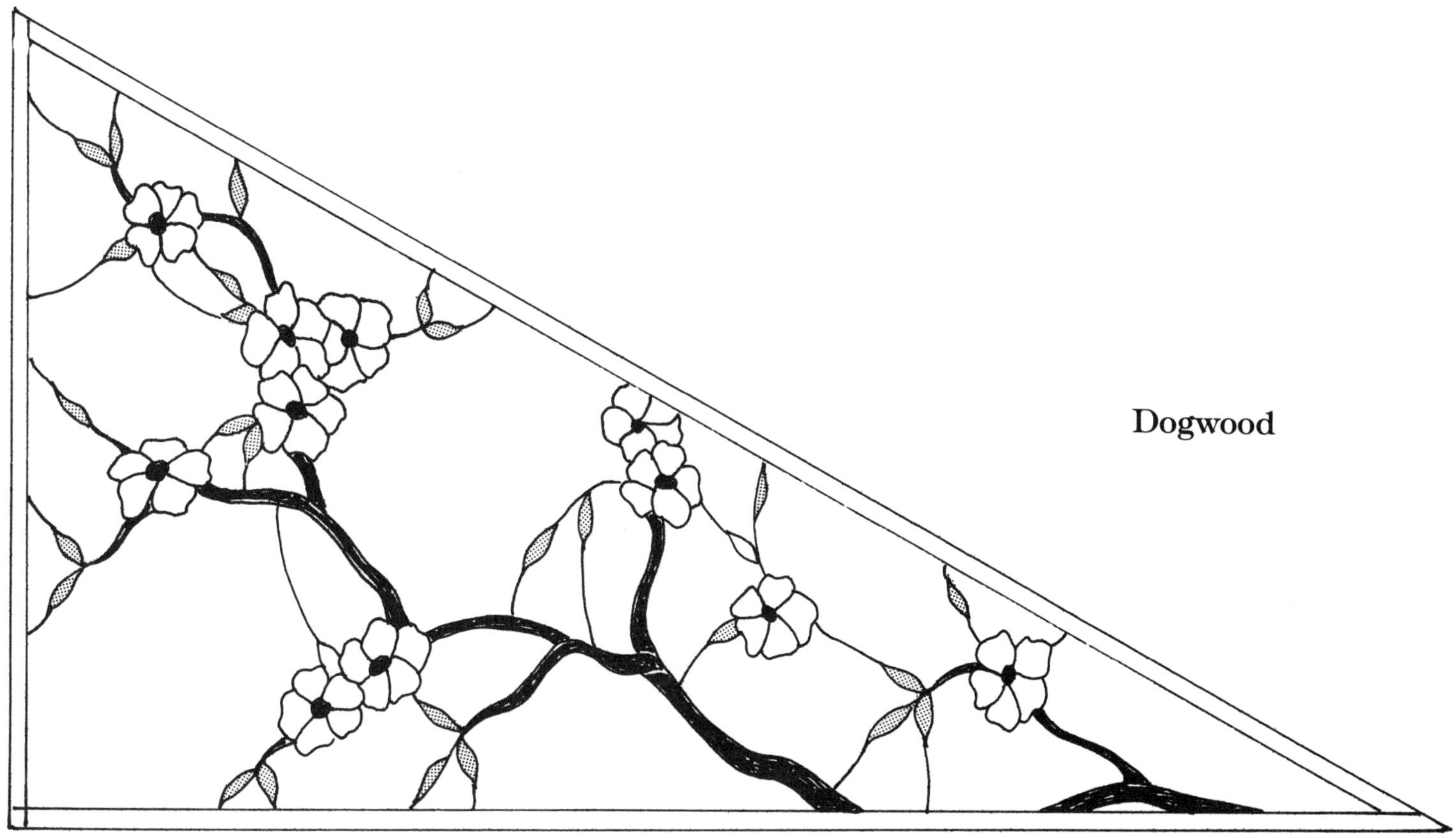

Dogwood

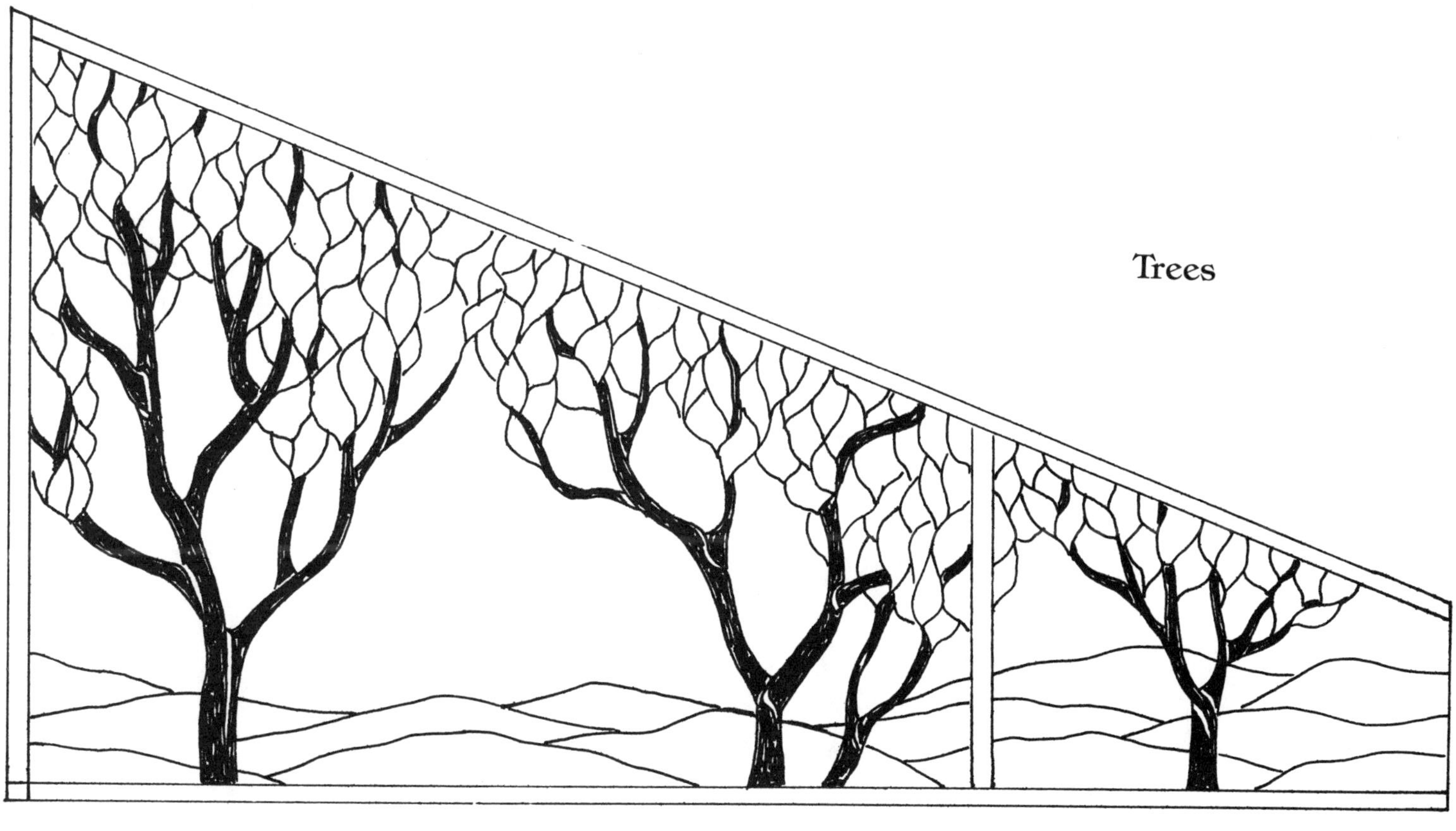

Trees

A Canopy of Dogwood

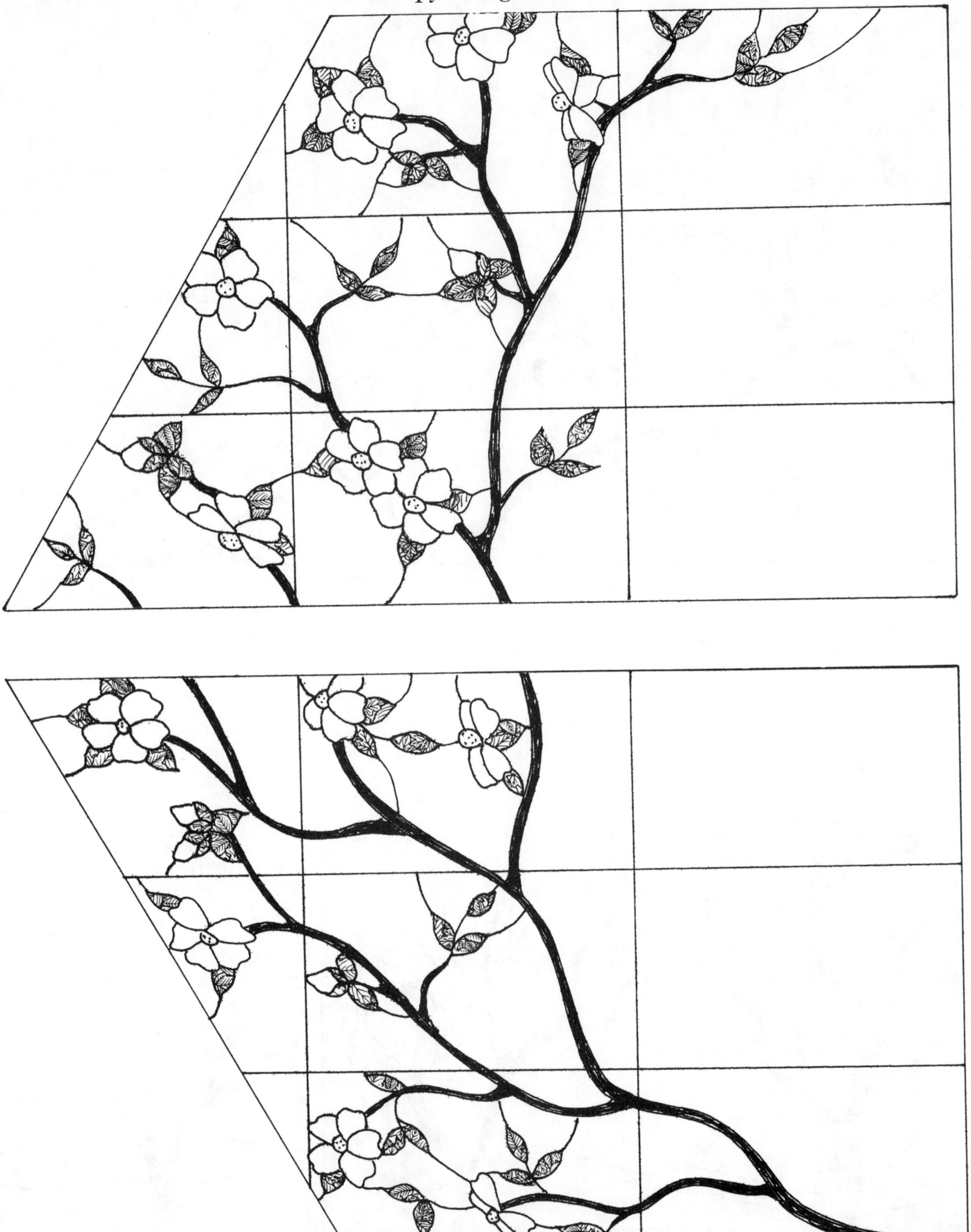

Butterfly Symphony

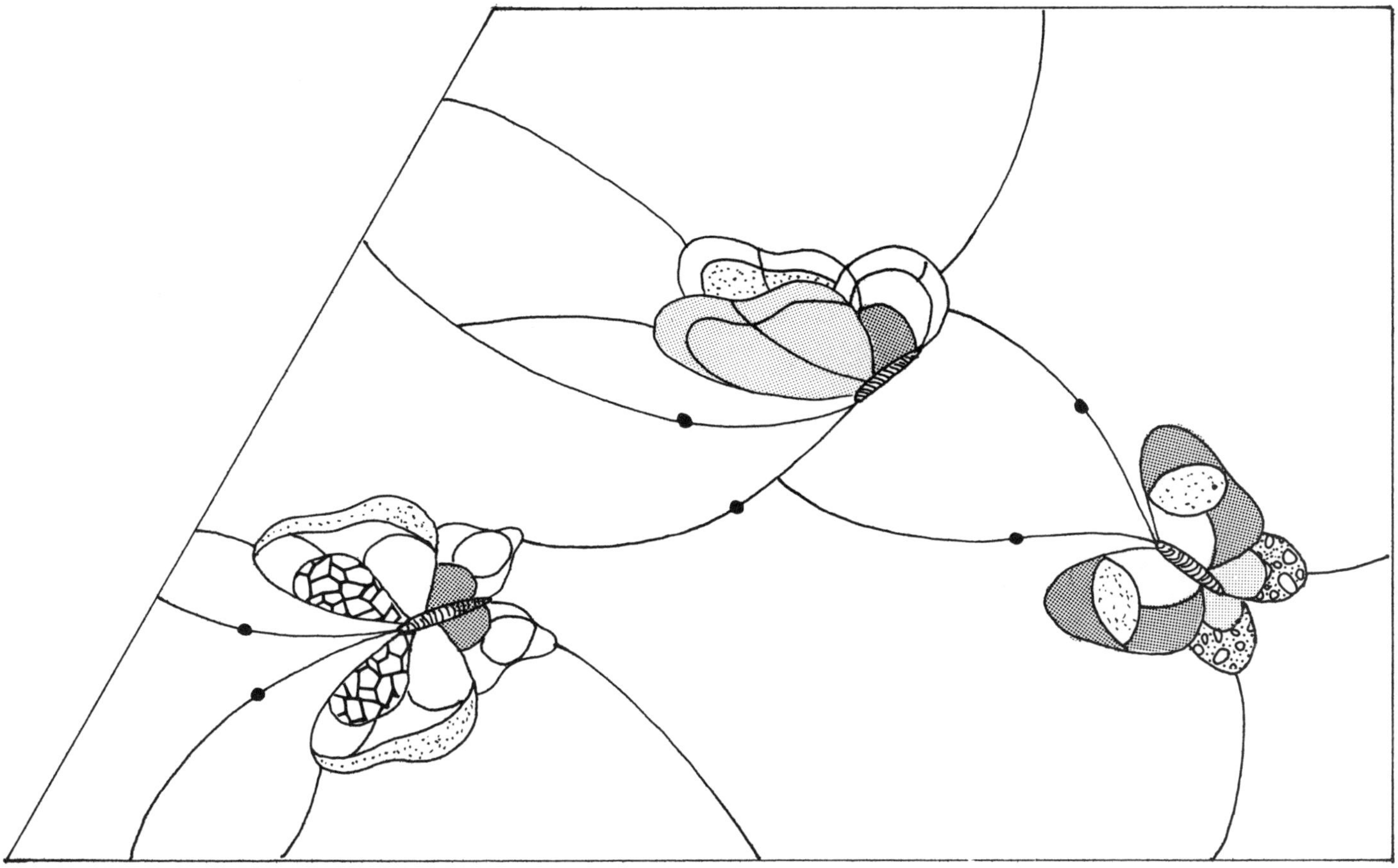

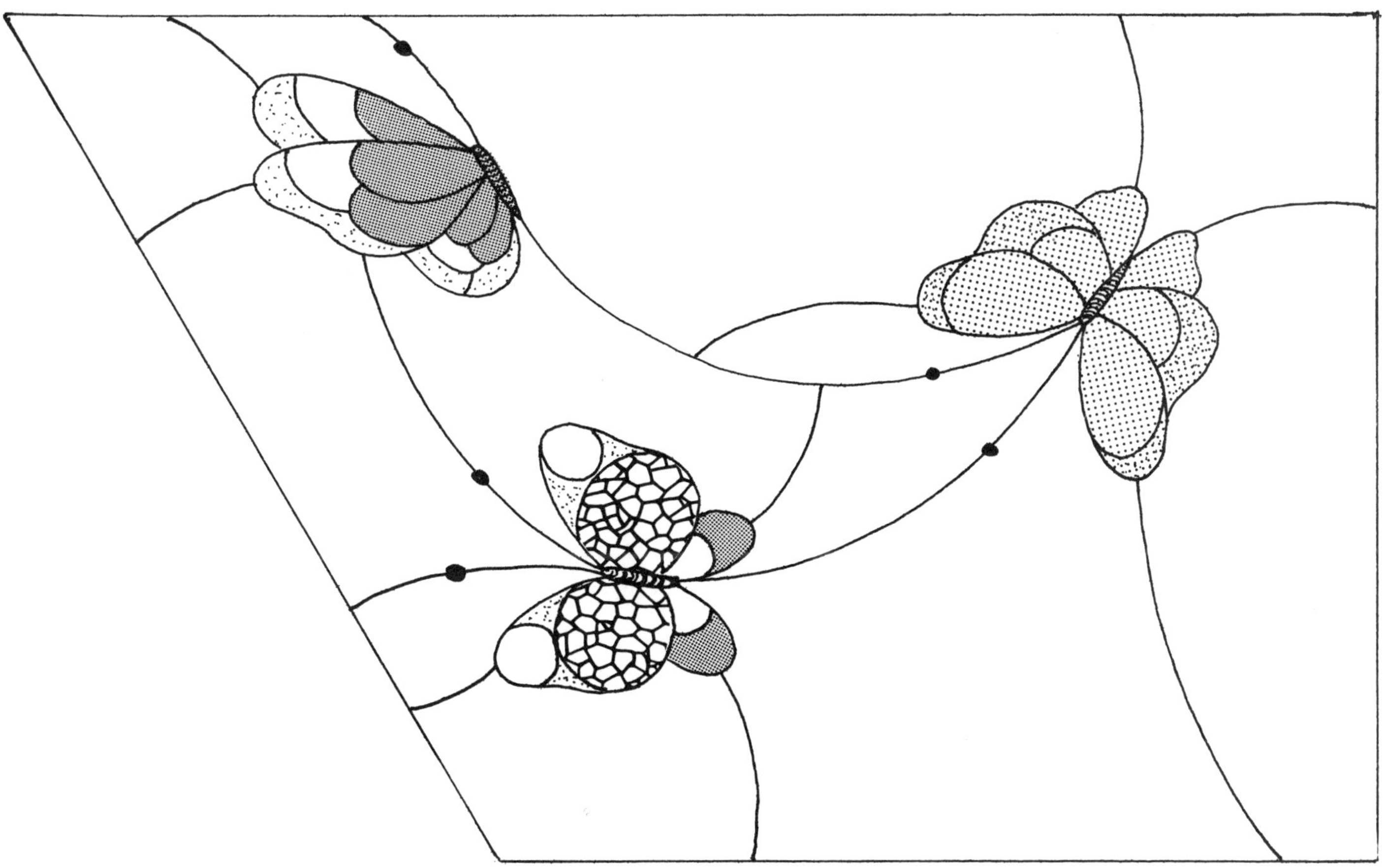

Wild Roses

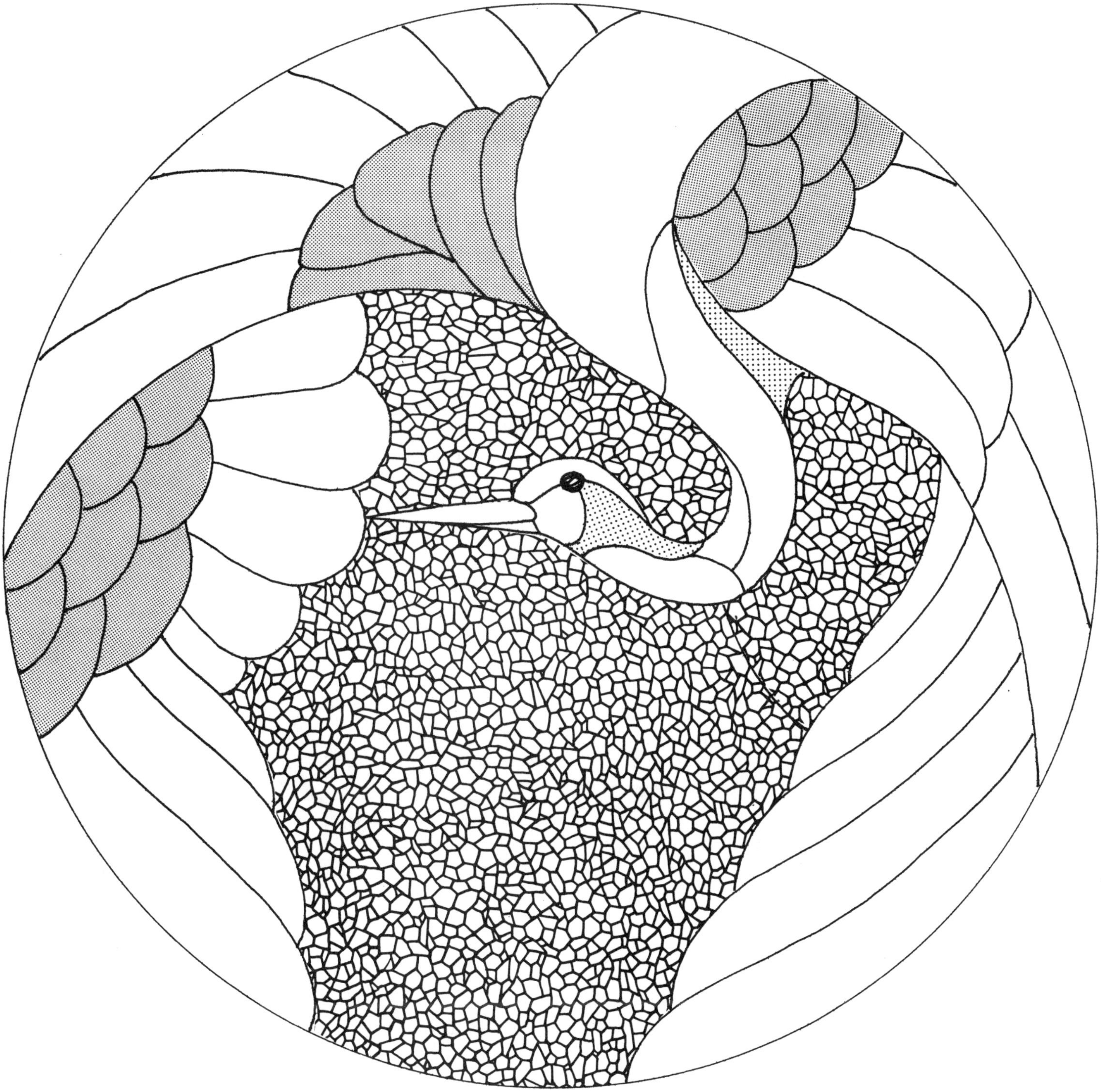

This design can be easily modified to fit a square, rectangle or used as a medallion in a larger window.

Orchids

SCREENS AND ROOM DIVIDERS are wonderful in stained glass. Using the natural light during the day, and lamp lighting at night keeps these stained glass pieces alive and exciting all the time. Including bevels in these pieces adds a quality that enhances and enriches the screens.

Individual screen panels can be used as sidelights enlarged if necessary with the addition of a border.

Water Iris

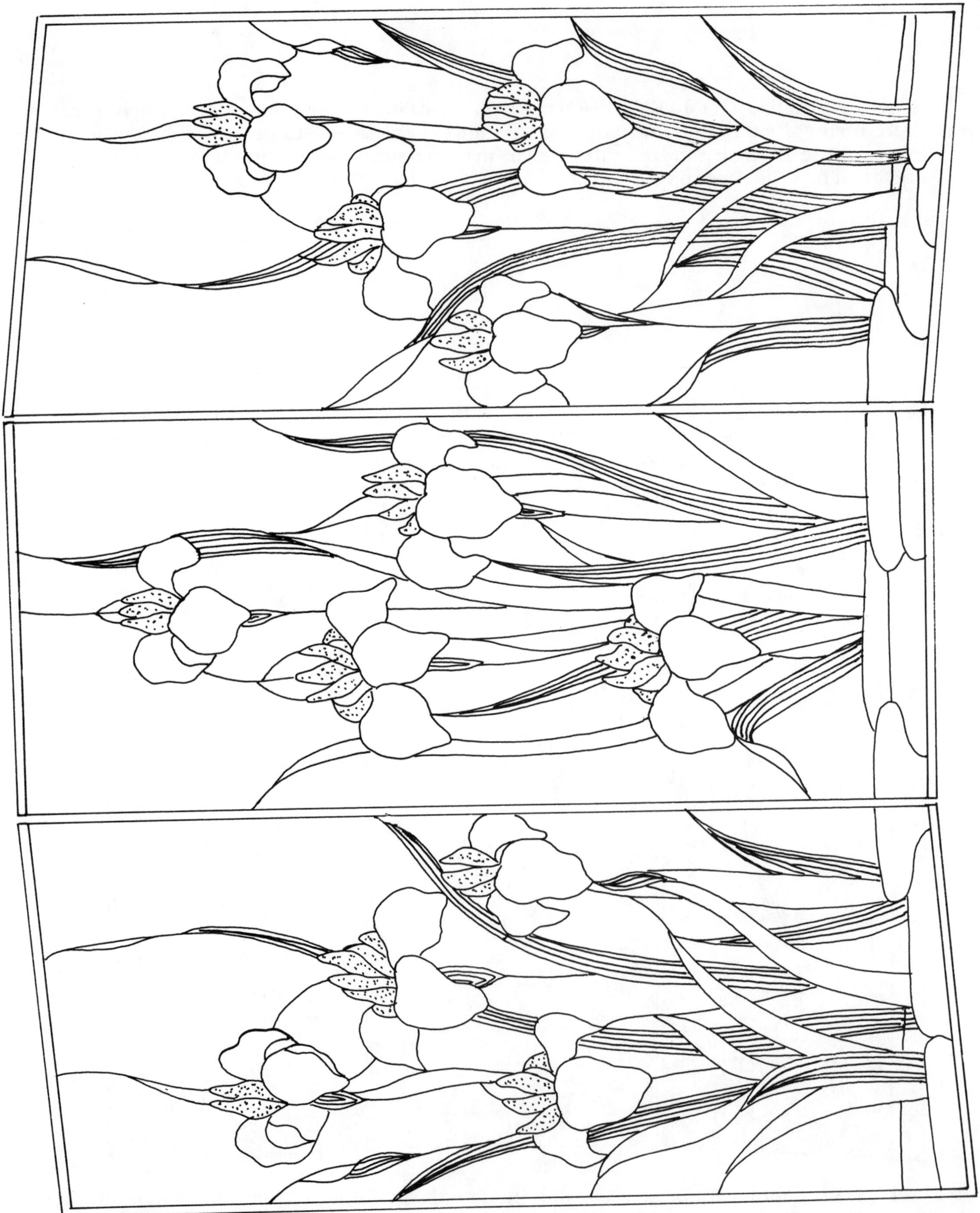

Lotus

Bamboo and Peonies

KITCHENS, BREAKFAST ROOMS AND DINING ROOMS are wonderful areas for stained glass. Consider remodeling a section of your cabinet doors to accommodate stained glass. If opaque or semi-transparent glass is used you need not worry about keeping the cabinet contents in order. With the proper selection of glass it is not necessary to backlight the doors.

A series of fruits and vegetables is a delightful design idea. Flowers and birds are always good ideas. Geometric designs are another option. If you have a skylight in your kitchen—why not design a stained glass piece for it. Be sure to check your local building code on skylights. A tempered glass supporting the stained glass piece may be required.

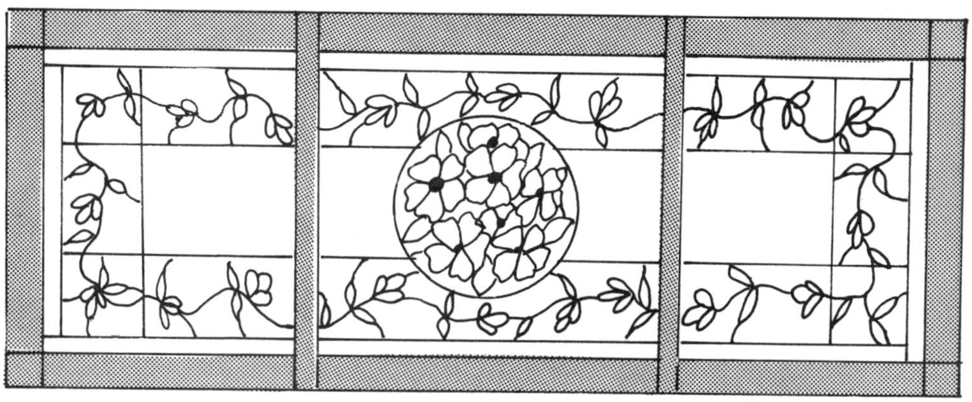

Designs for cabinet doors or individual windows

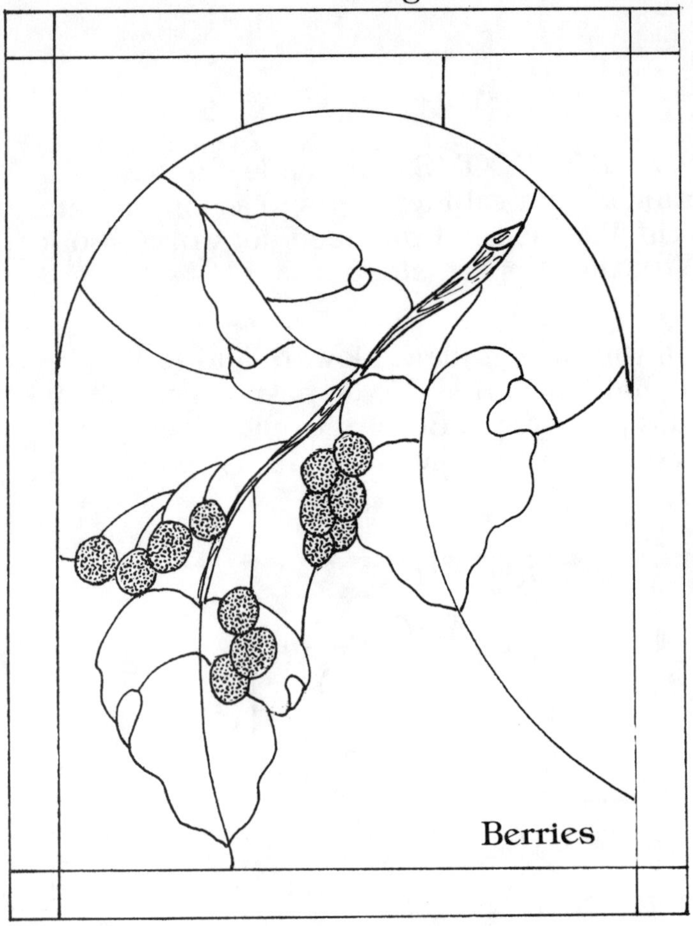

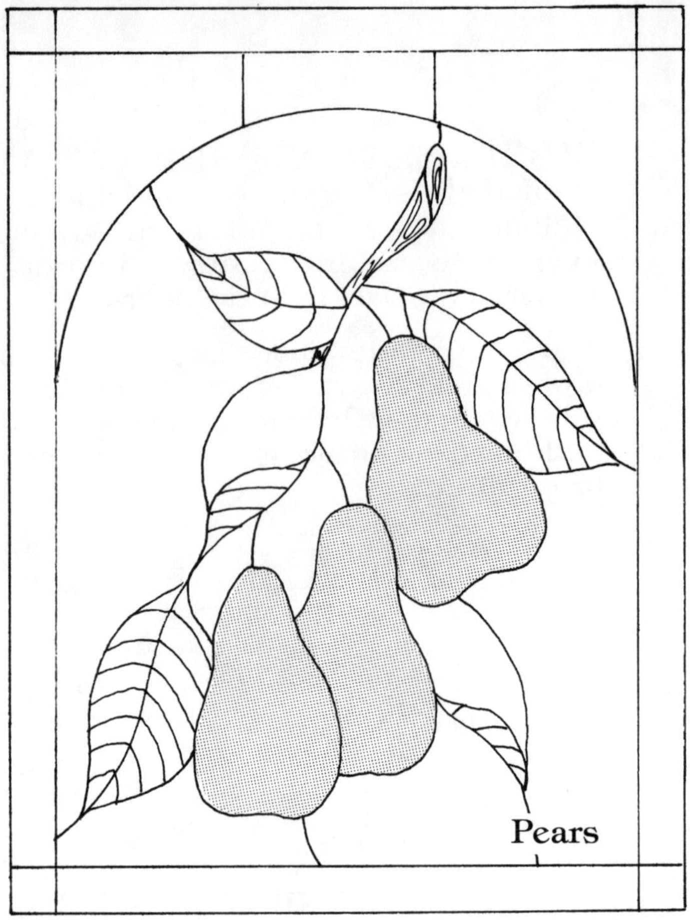

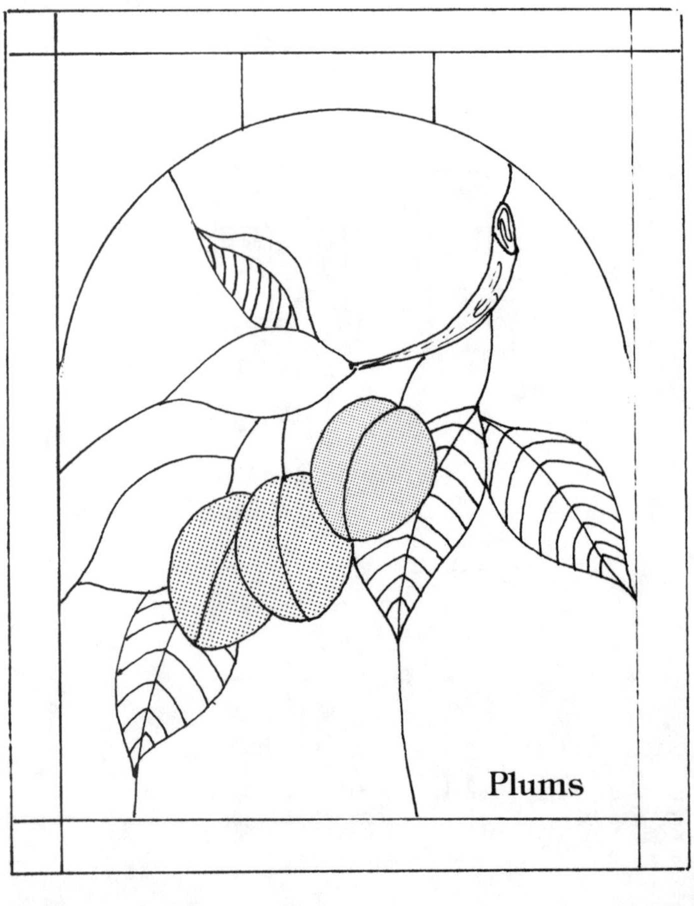

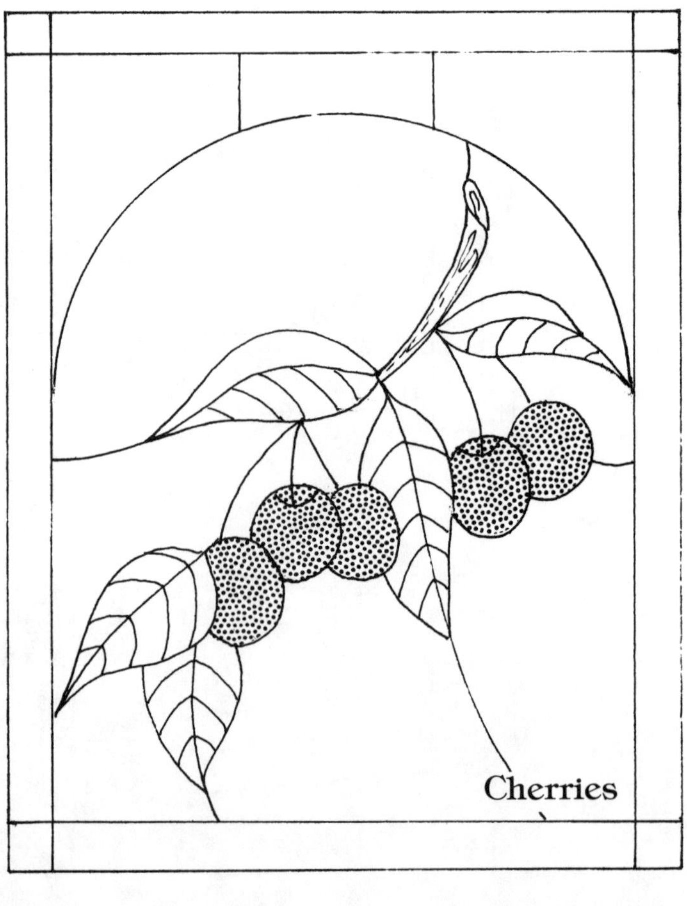

Apples

Oranges

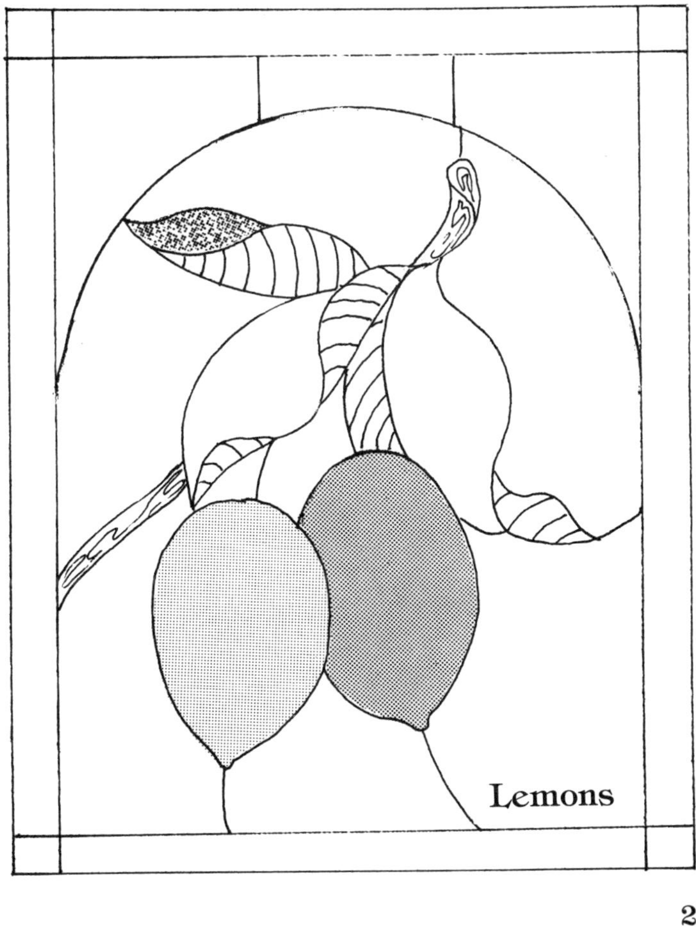
Lemons

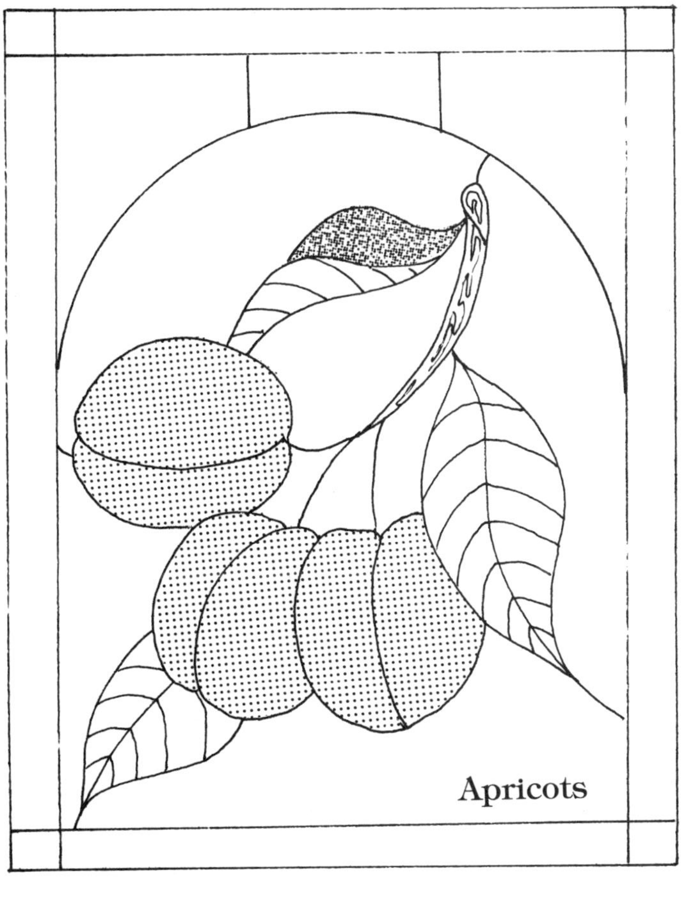
Apricots

Designs for cabinet doors or individual windows

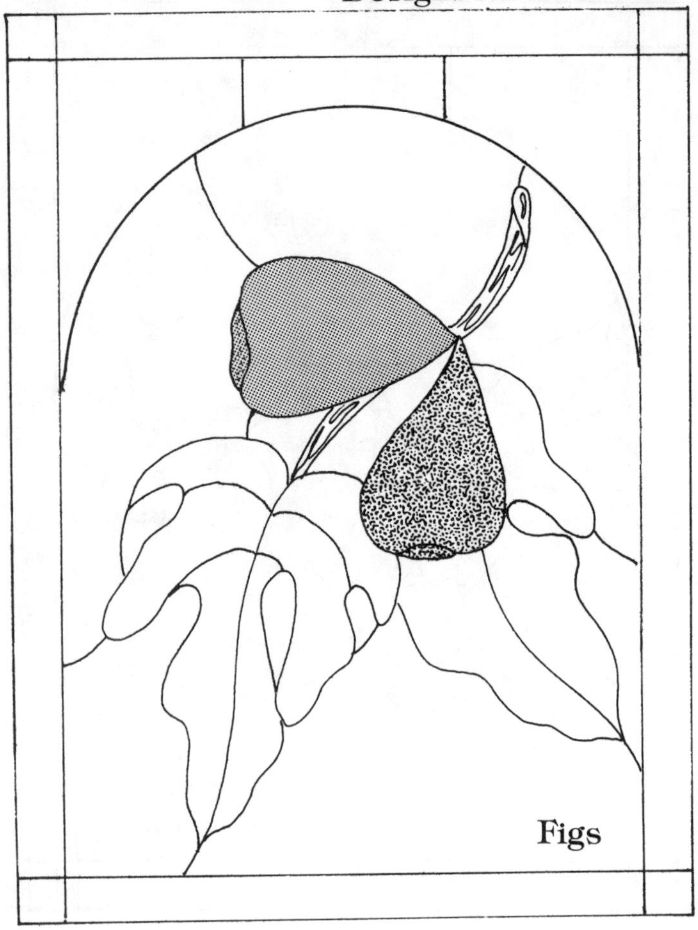

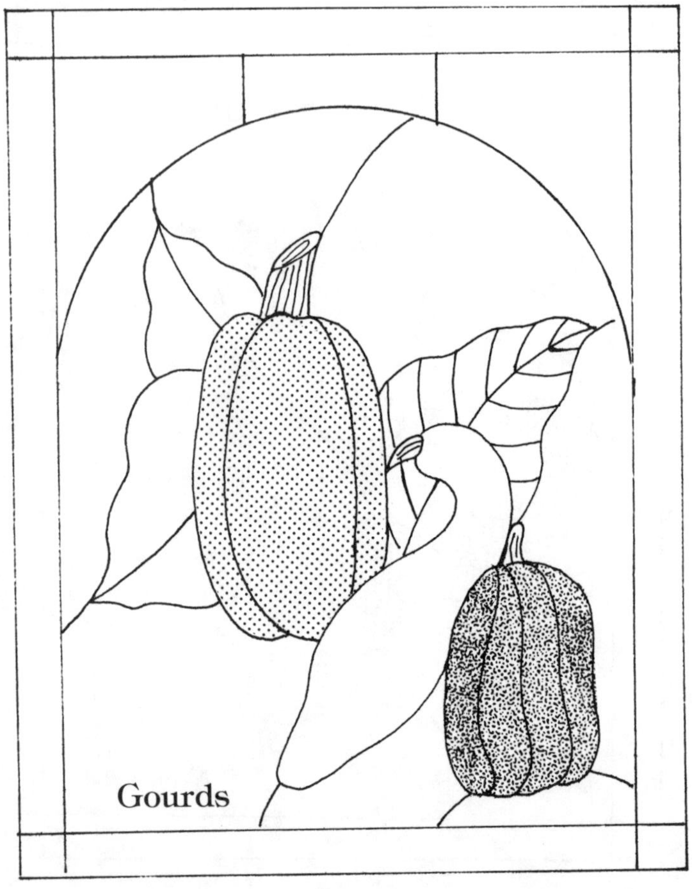

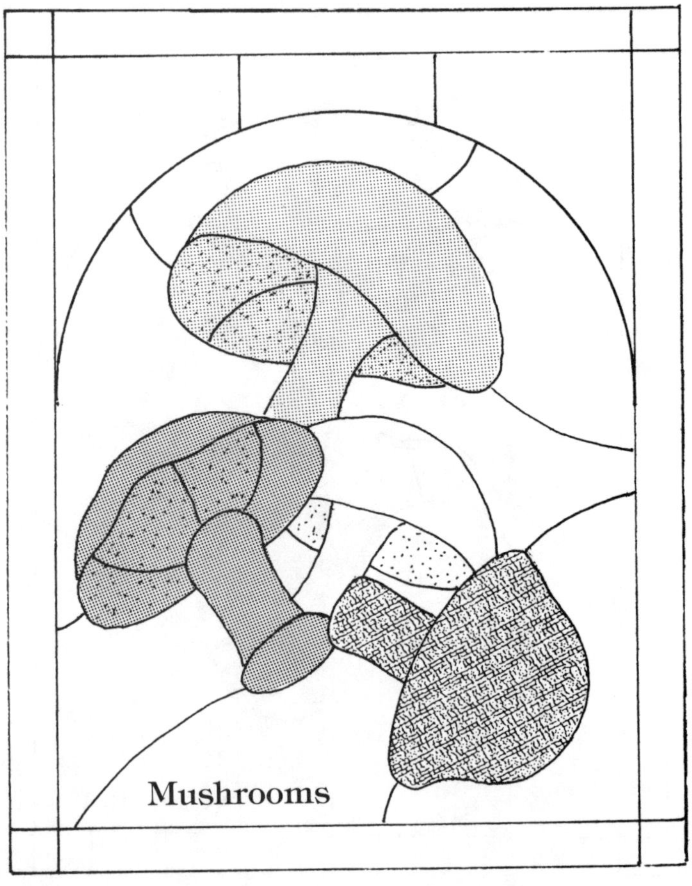

Mushrooms

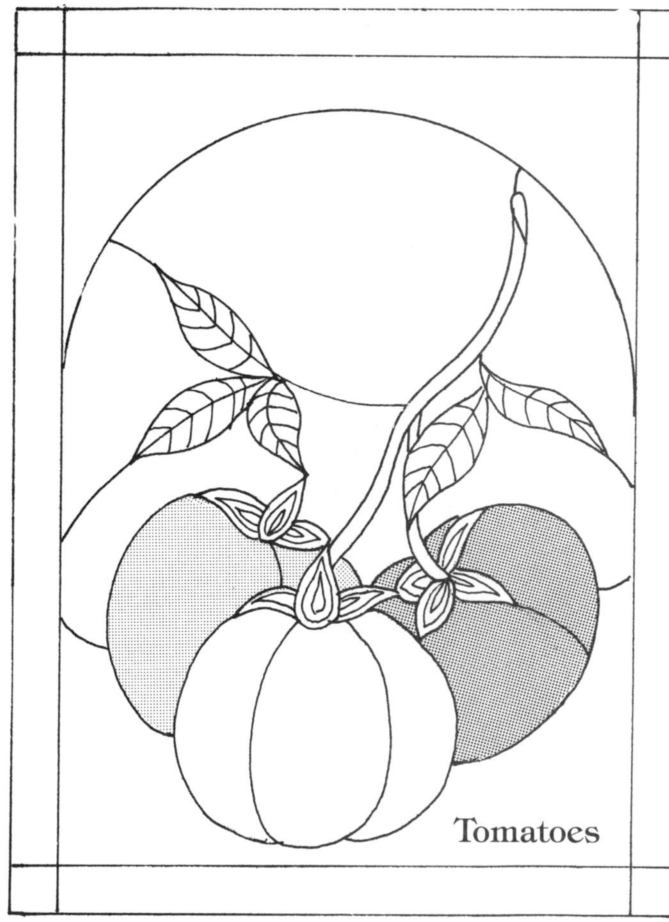
Tomatoes

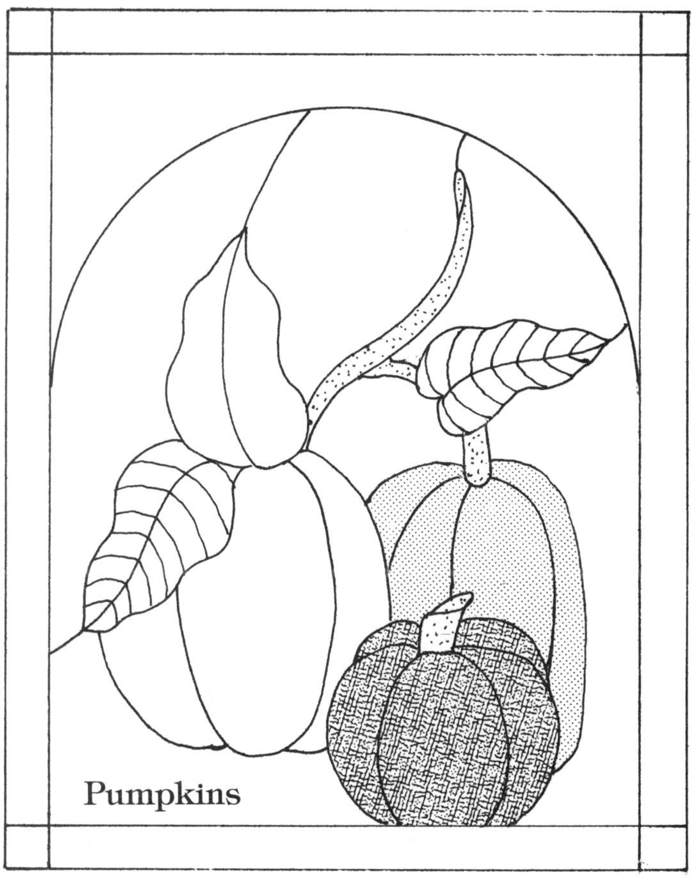
Pumpkins

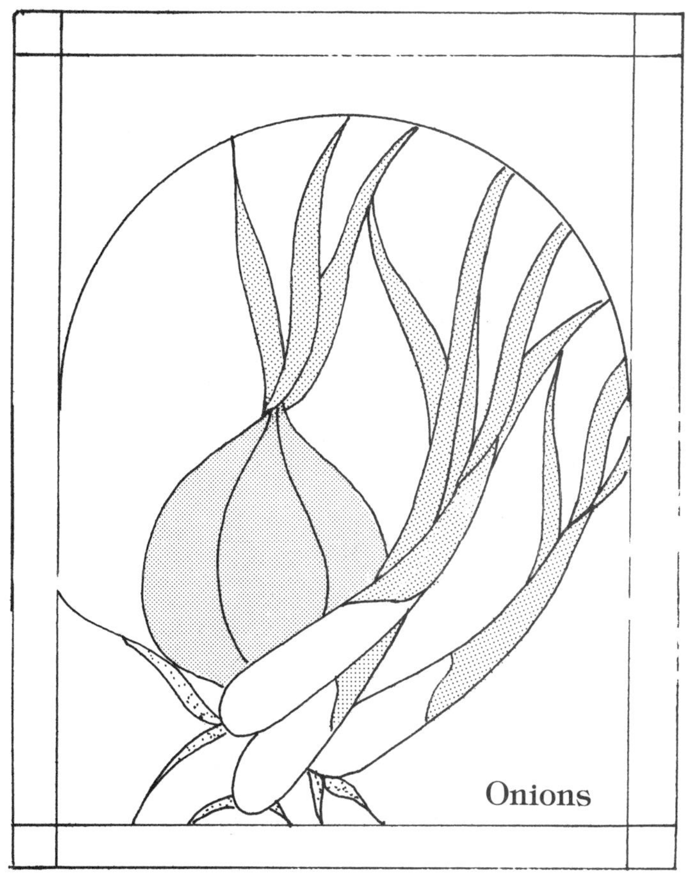
Onions

Designs for cabinet doors

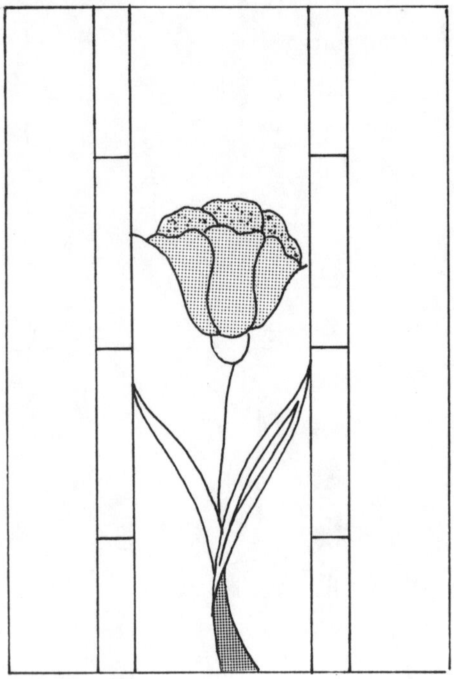
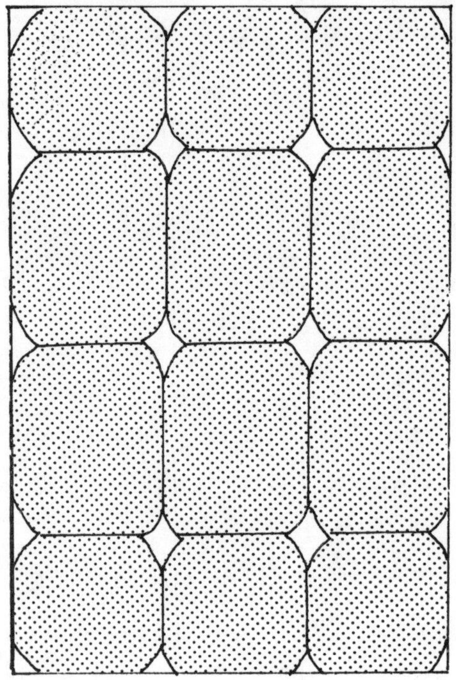
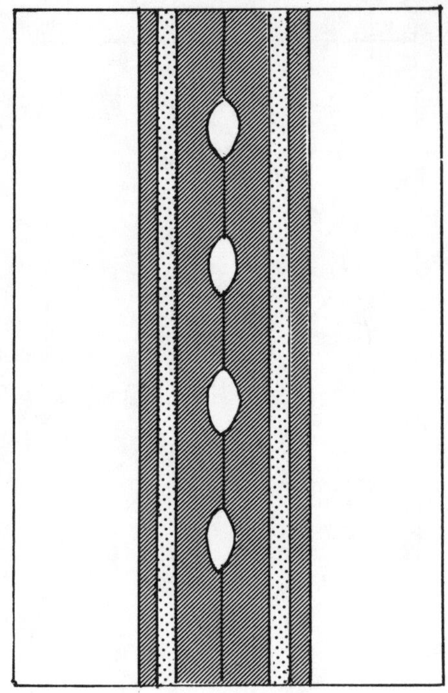
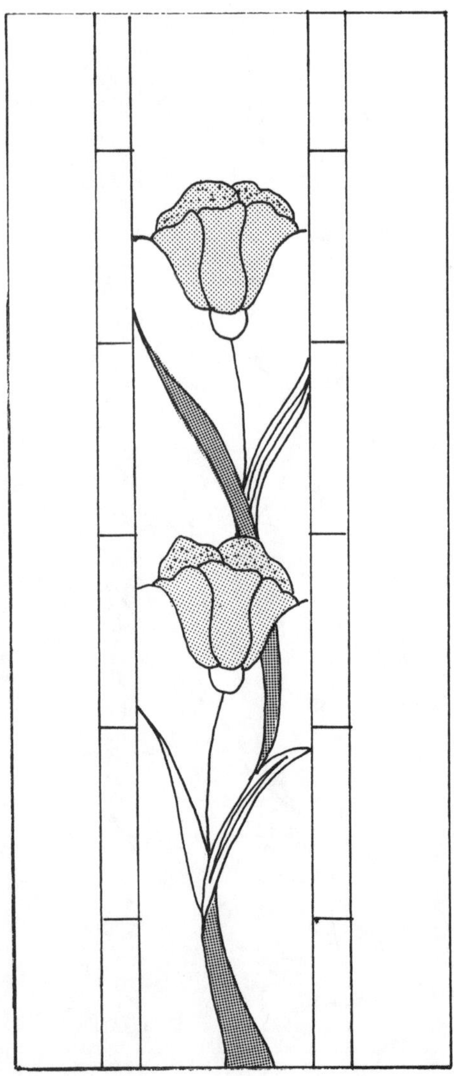
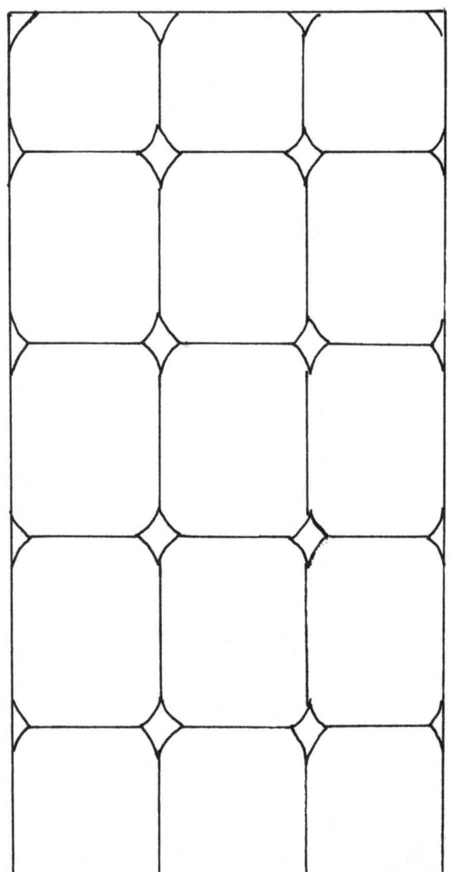
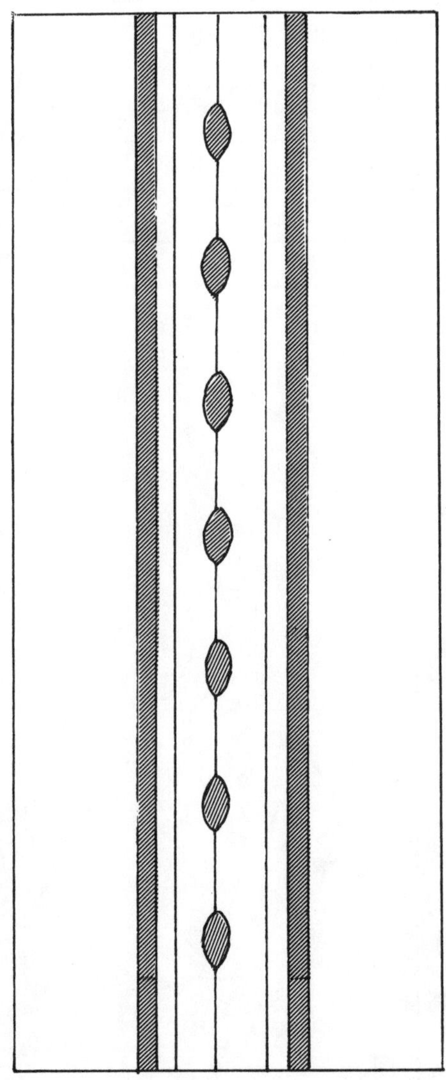

Designs for cabinet doors

Designs for cabinet doors

Designs for cabinet doors

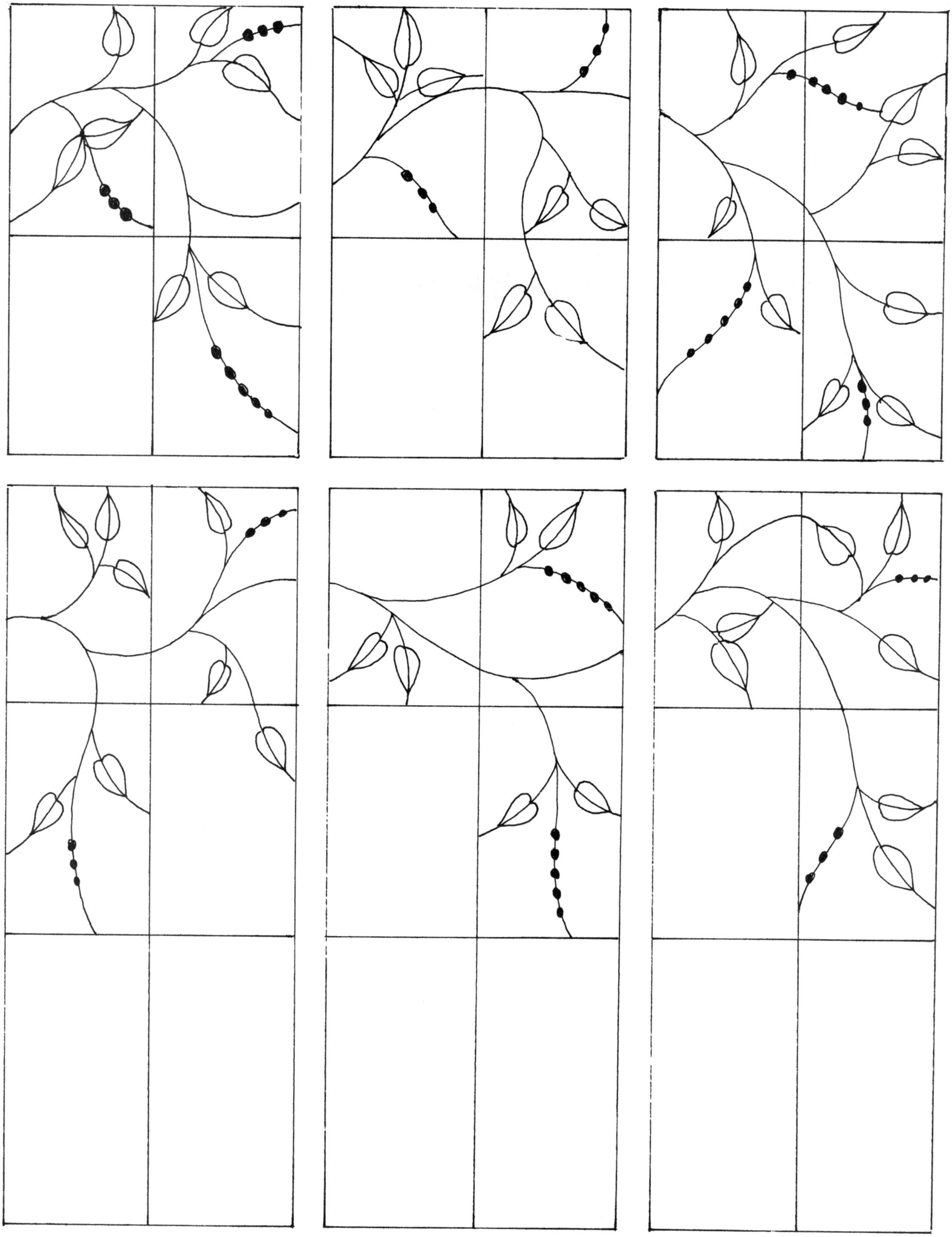

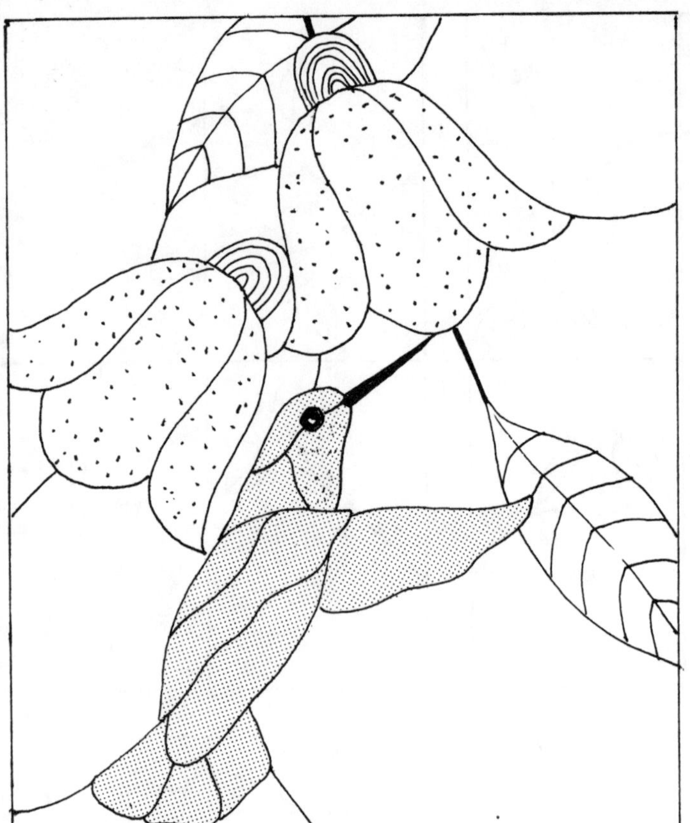

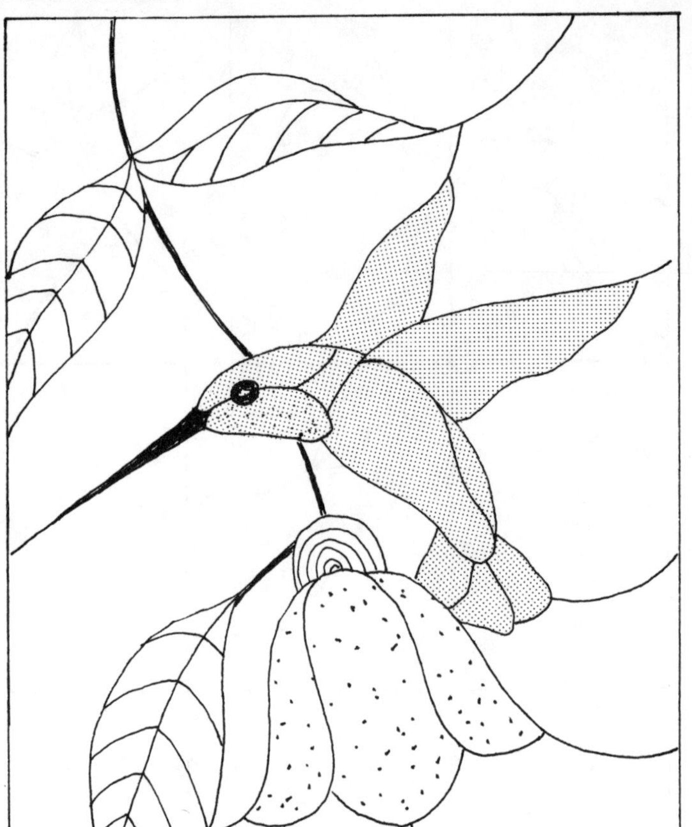

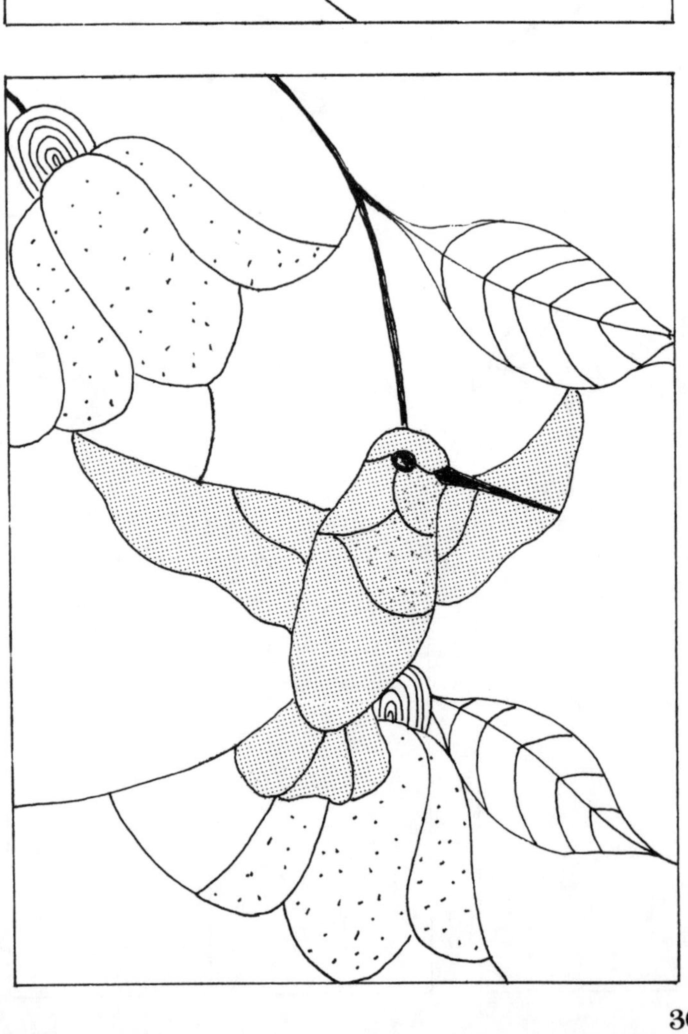

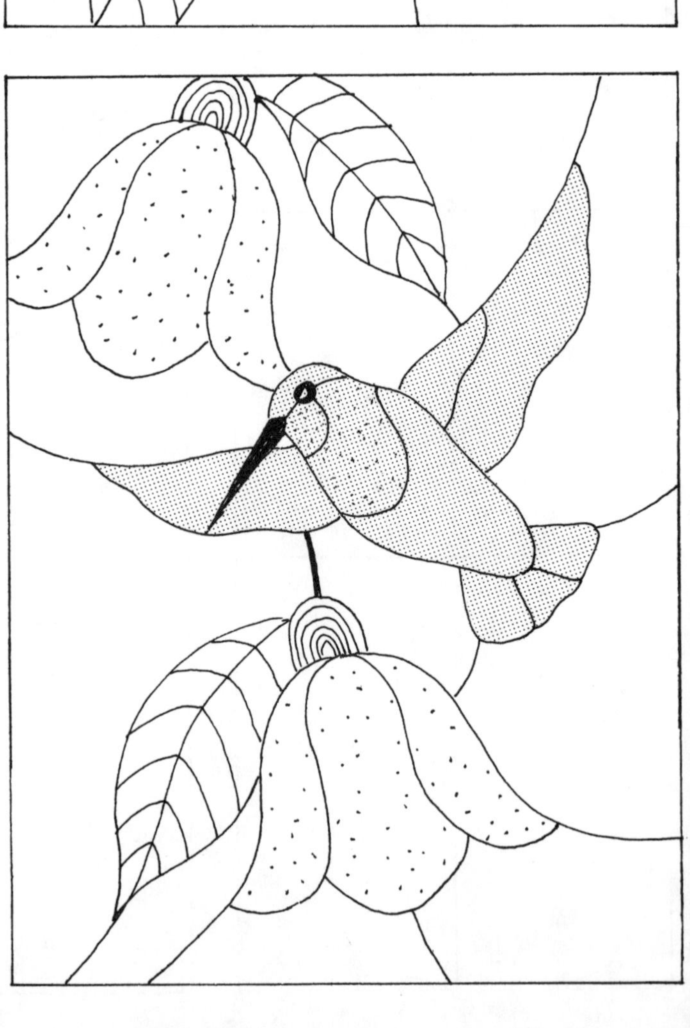

Designs for cabinet doors

Designs for cabinet doors

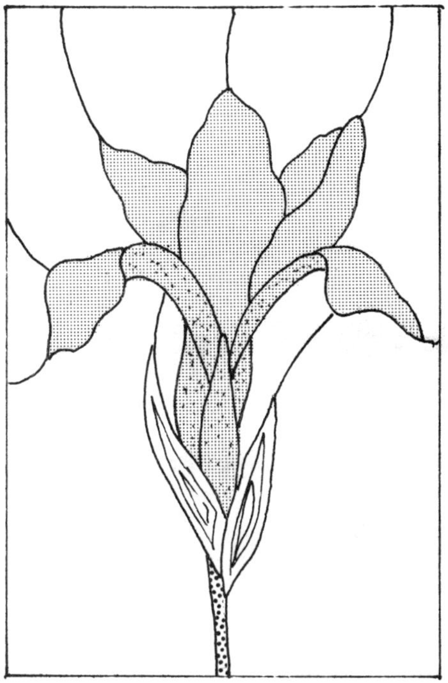

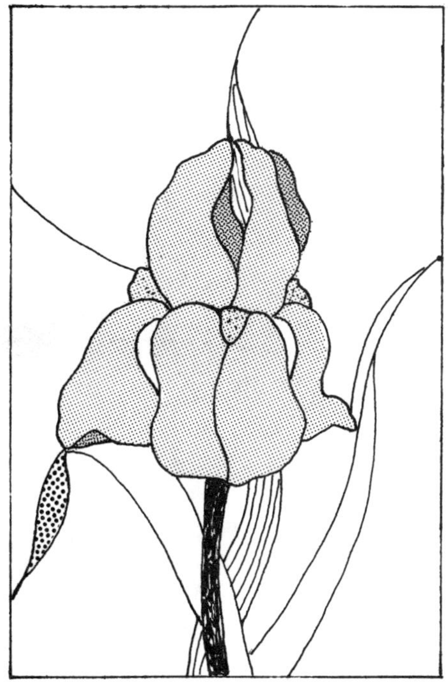

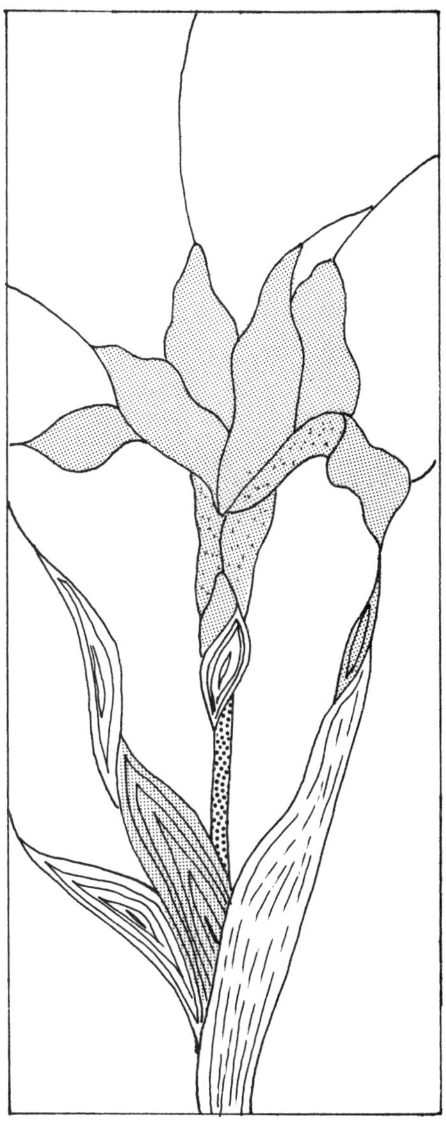

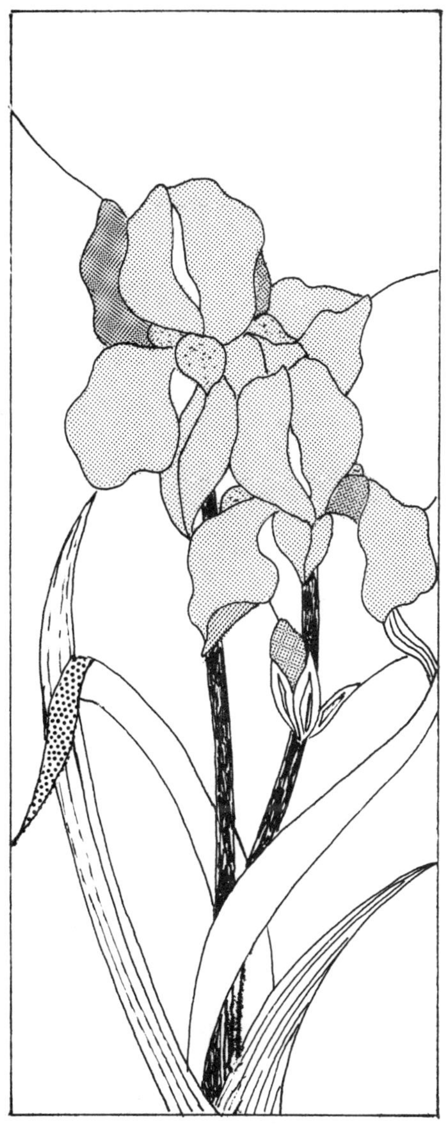

Designs for cabinet doors

Designs for cabinet doors

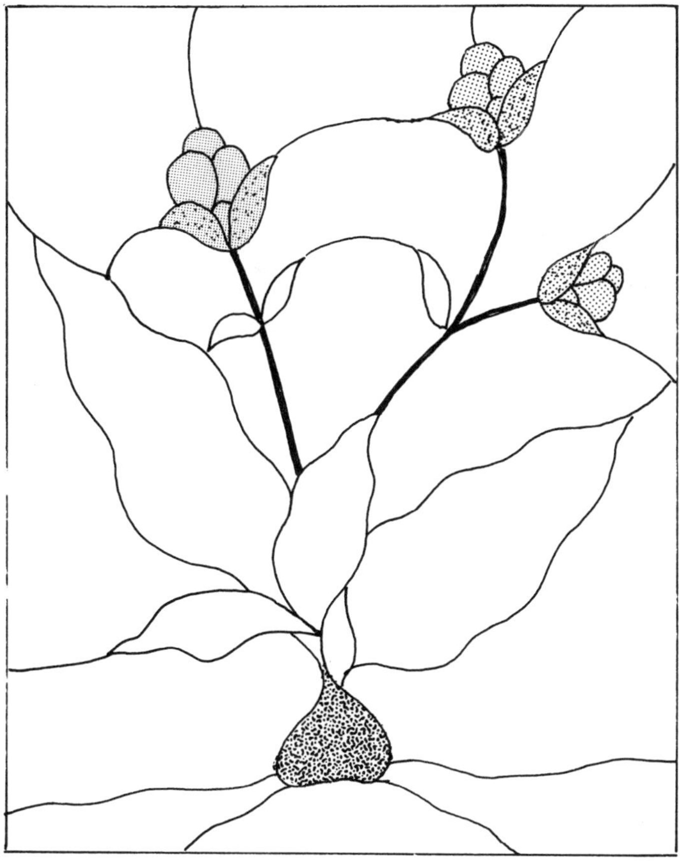

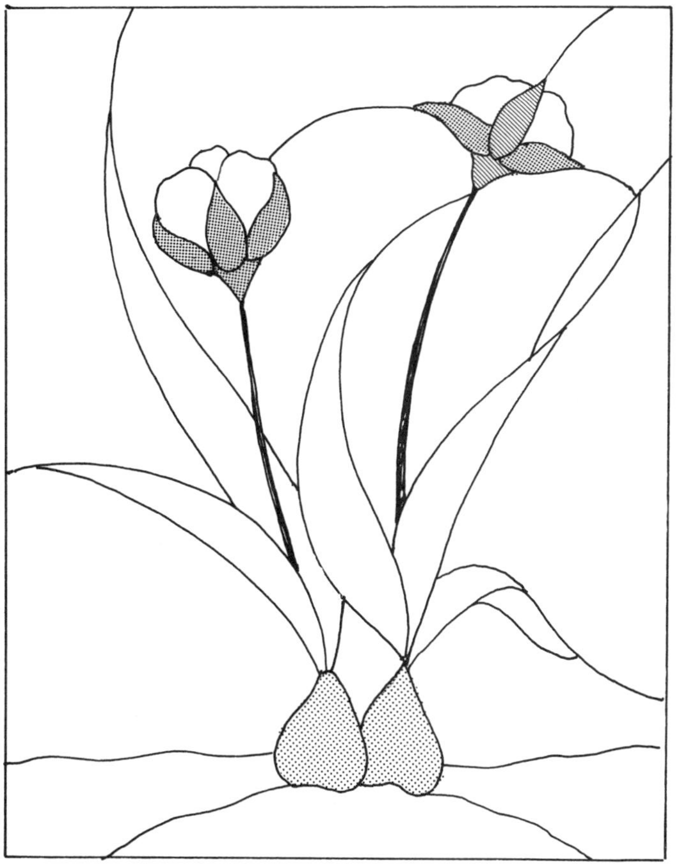

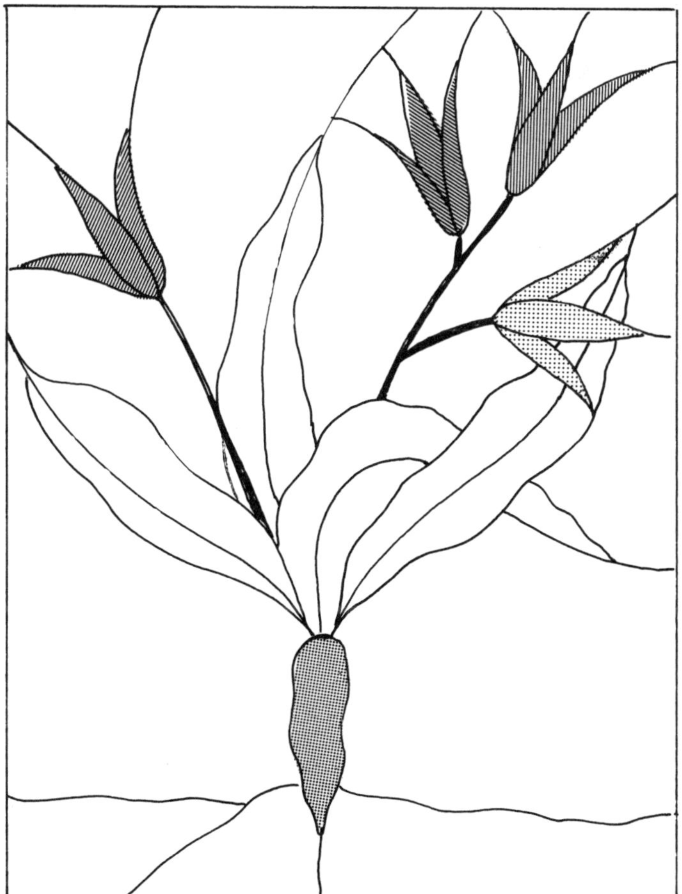

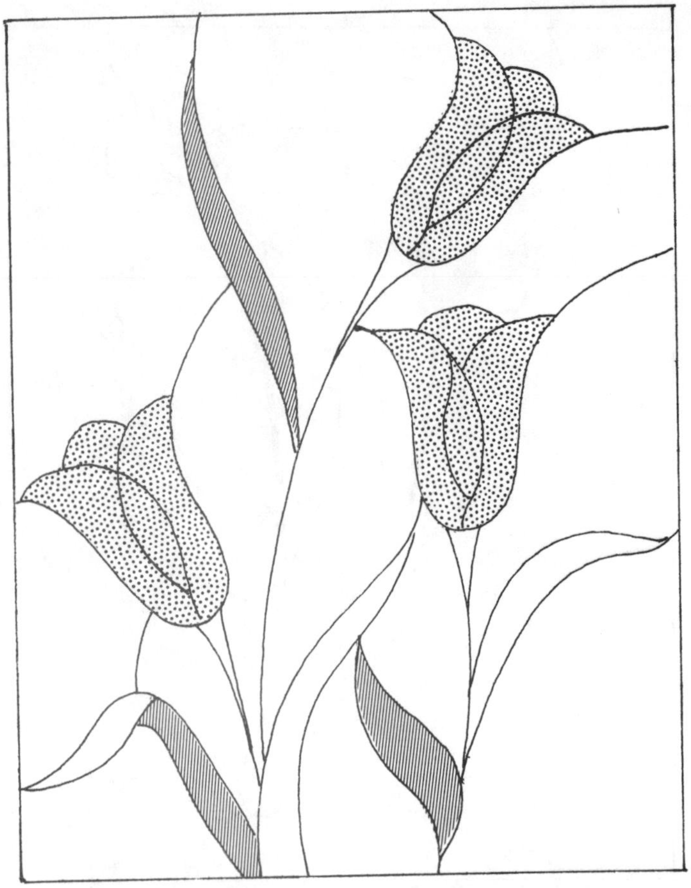
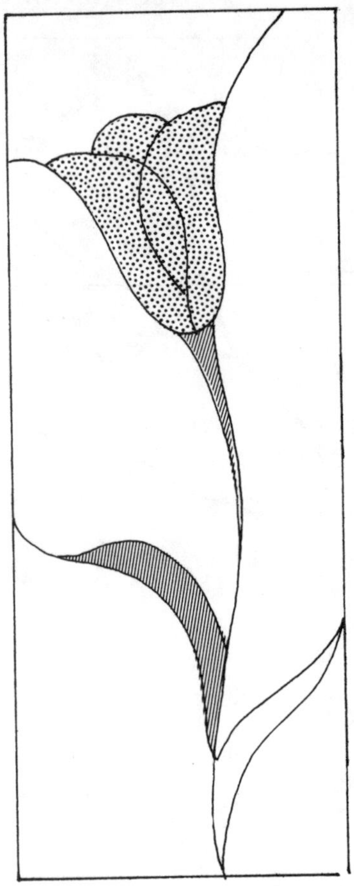
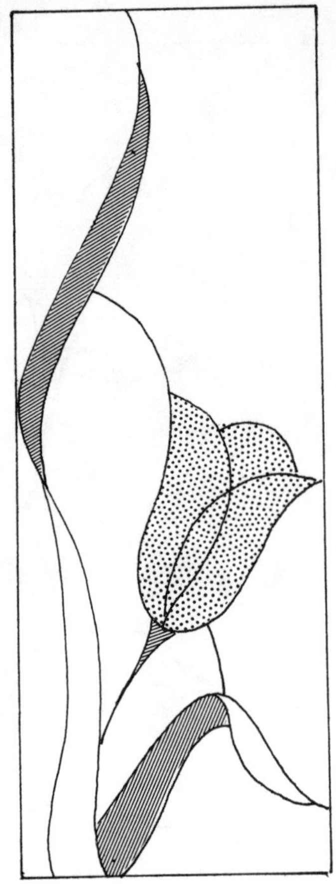

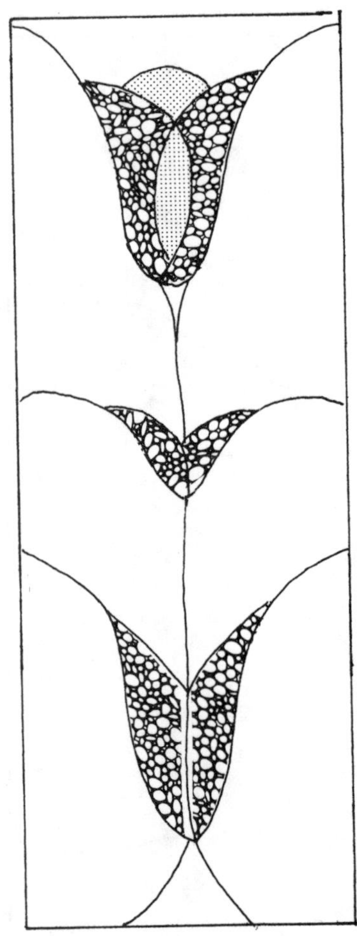
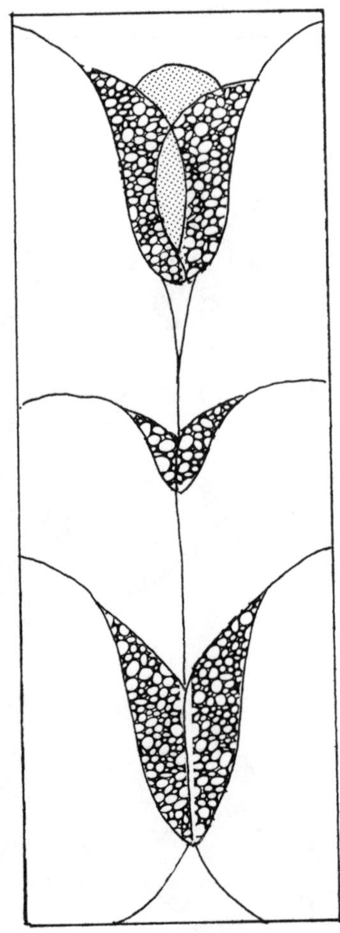

Wine and Orchids

Daylilies

Flowers with oval jewels

The Fountain

Vista

Chrysanthemums

The Oak

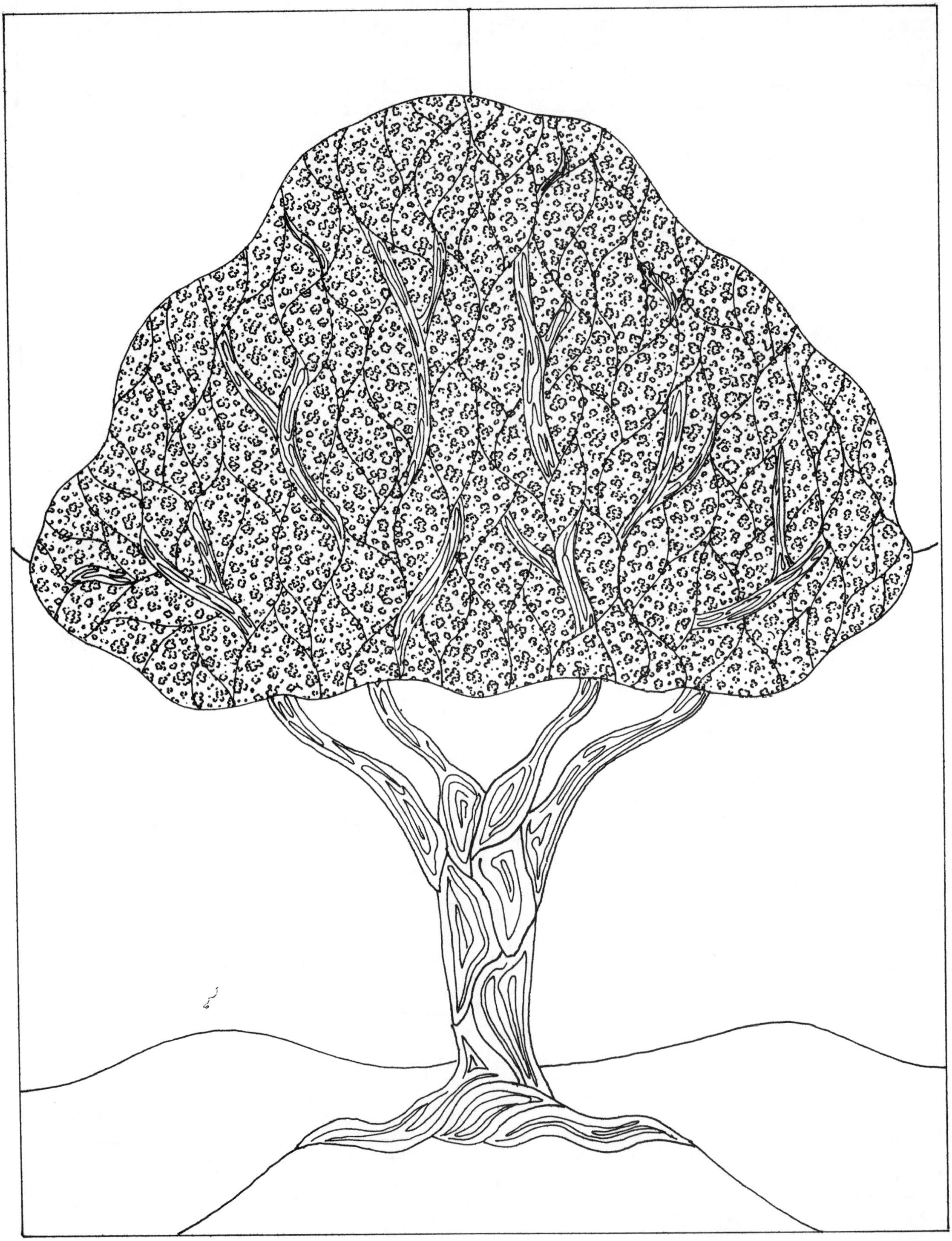

Monterey Cypress

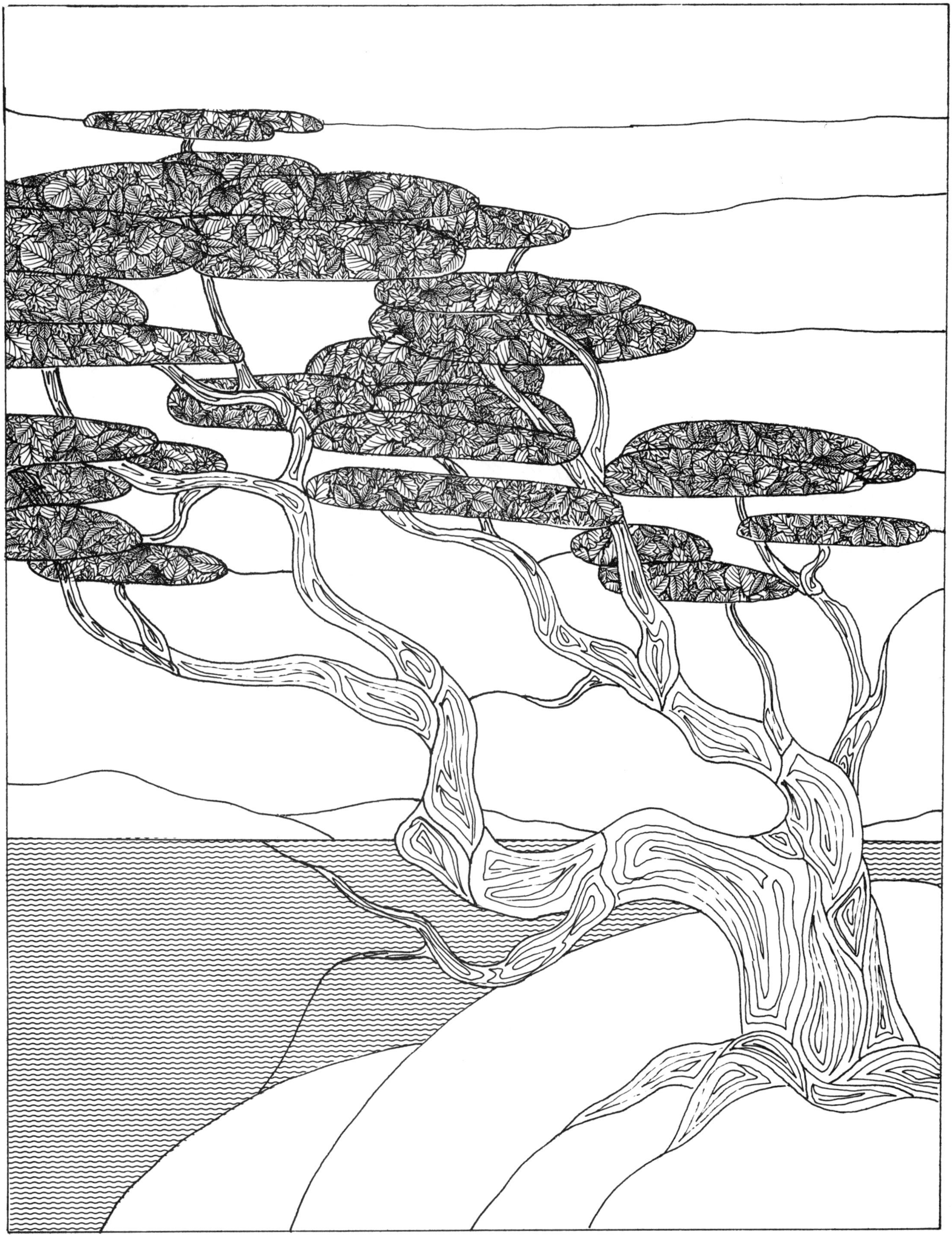

Bamboo

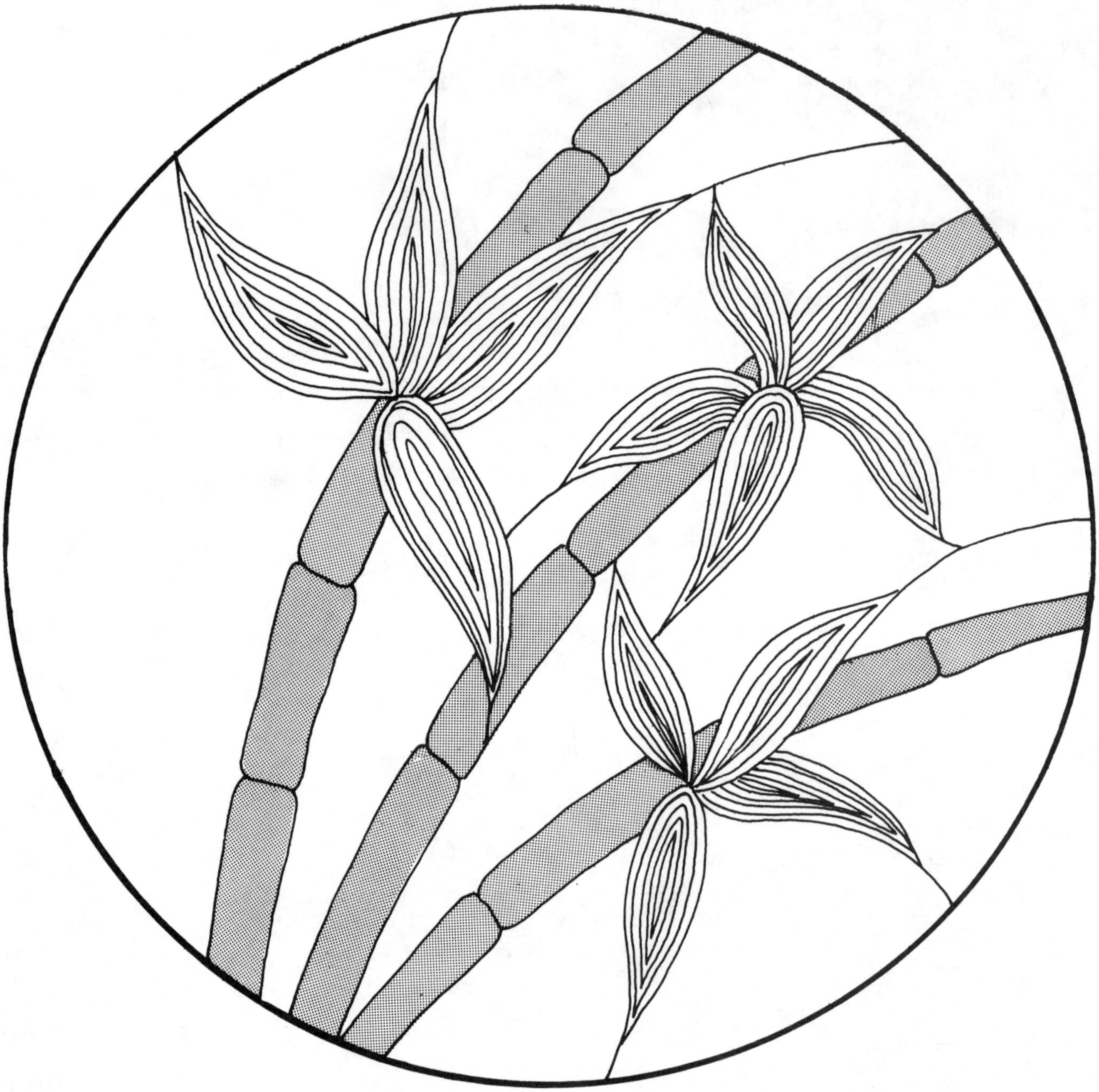

Daisies

– makes a wonderful patio table top

Flower Basket

– a beautiful room divider

FAMILY ROOMS AND GAME ROOMS offer great design possibilities—anything goes. Your favorite animals, hobbies, a "dream" can all be recaptured in glass. An entire mood of a room can be determined by the stained glass window or windows. Allow your imagination to soar in these rooms.

If the room has a fireplace—how about a stained glass firescreen! The screen is not only decorative but will be energy effective as well. Tempered glass behind the stained glass piece will distribute the heat better. Keep the screen 6 to 8 inches away from a roaring fire. When the fireplace is not in use the screen sits against the opening and prevents warm house air from escaping up the chimney.

Dolphins at Play

– an ideal table top design.

Birds in Flight

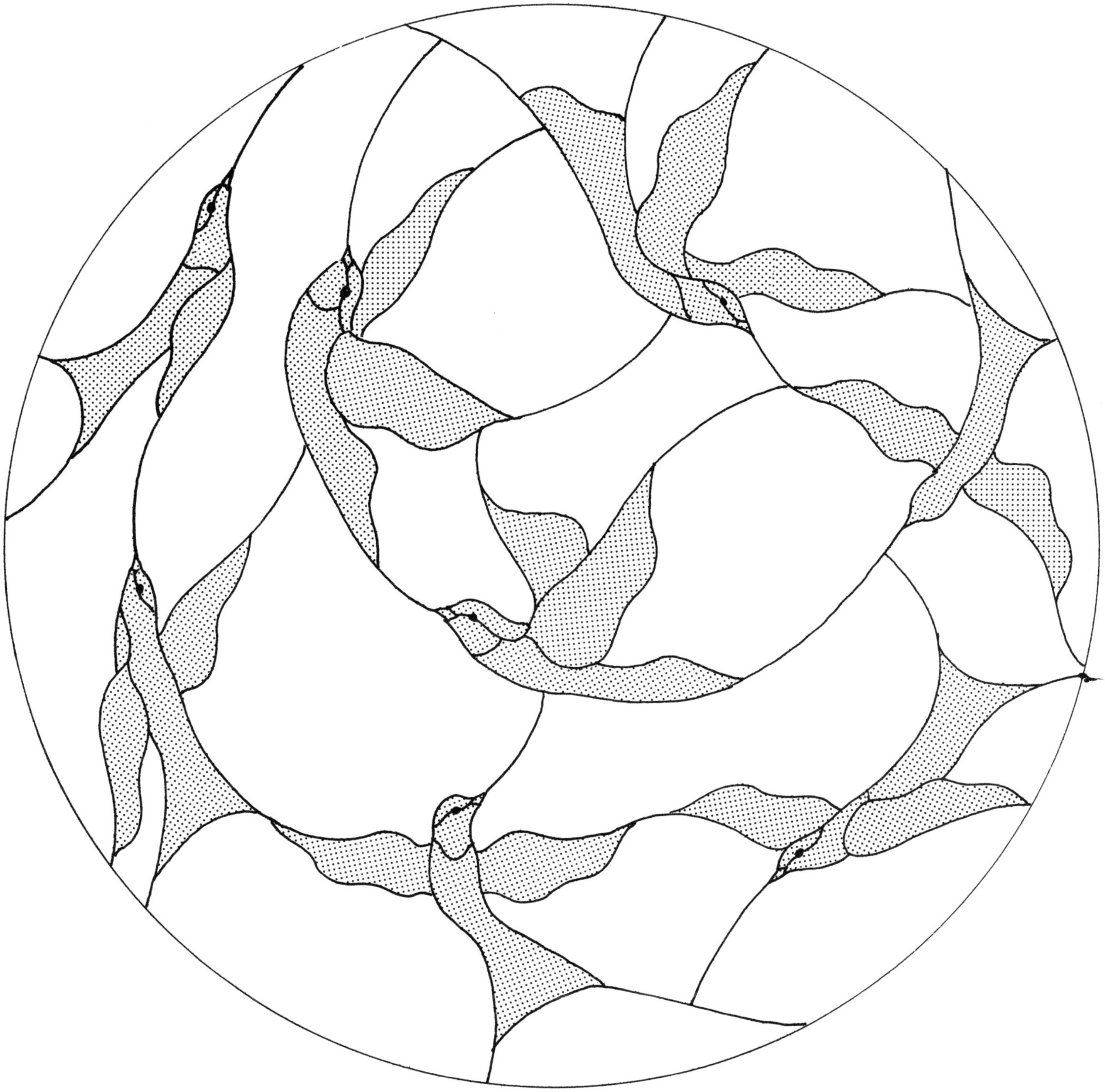

Shield

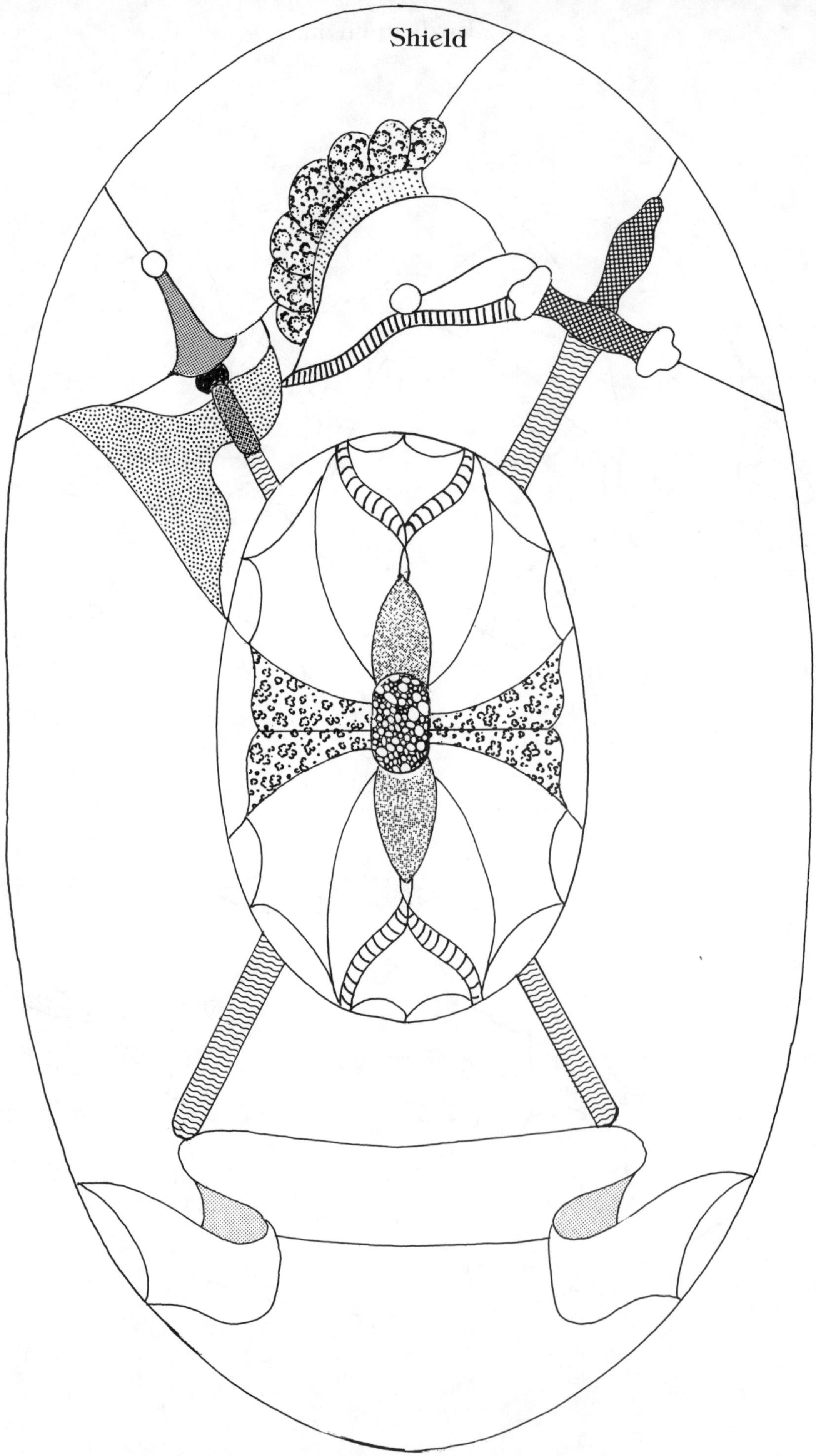

A Knight in Armor

A Carousel Horse

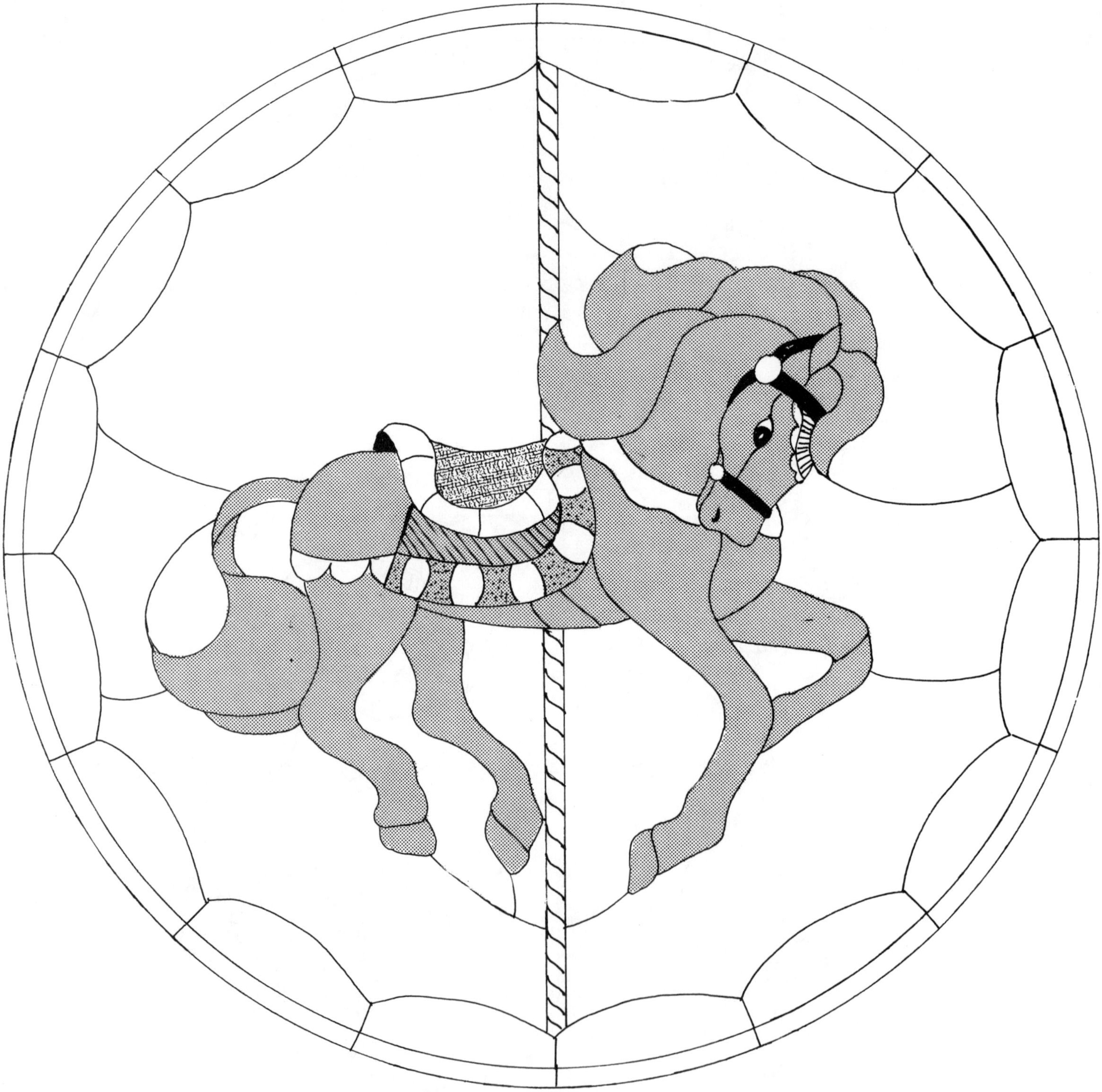

A Carrousel Goat

A Circus Horse

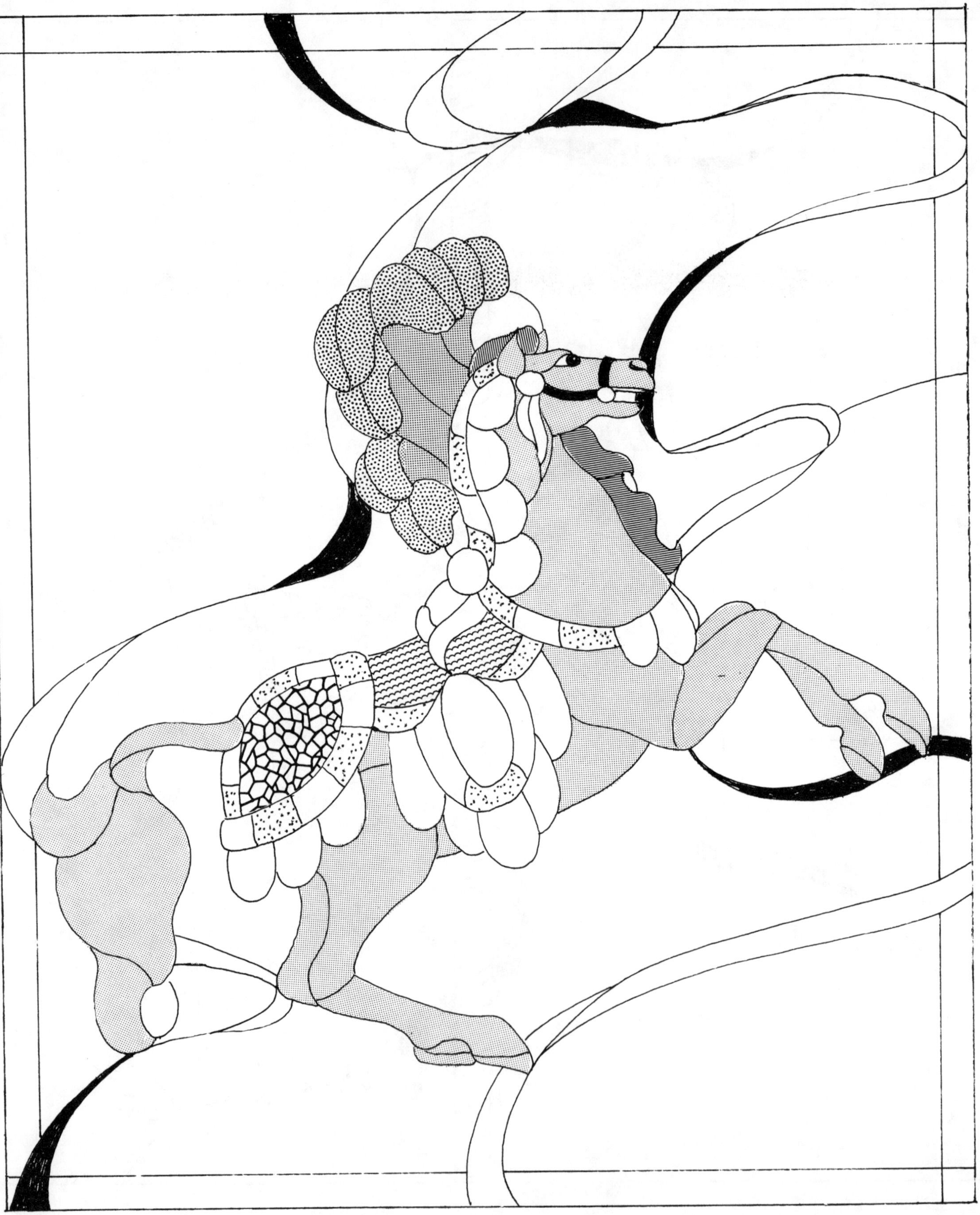

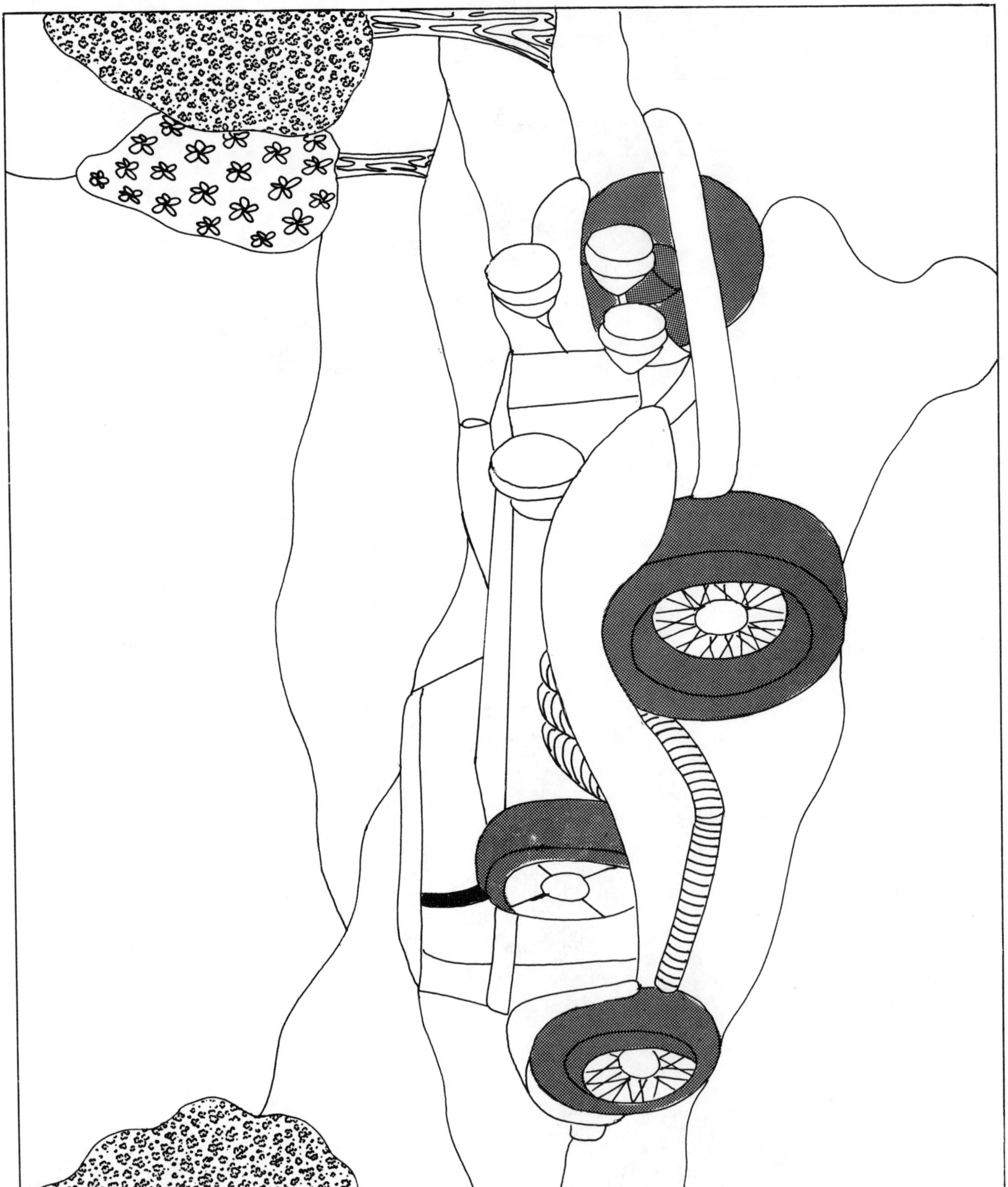

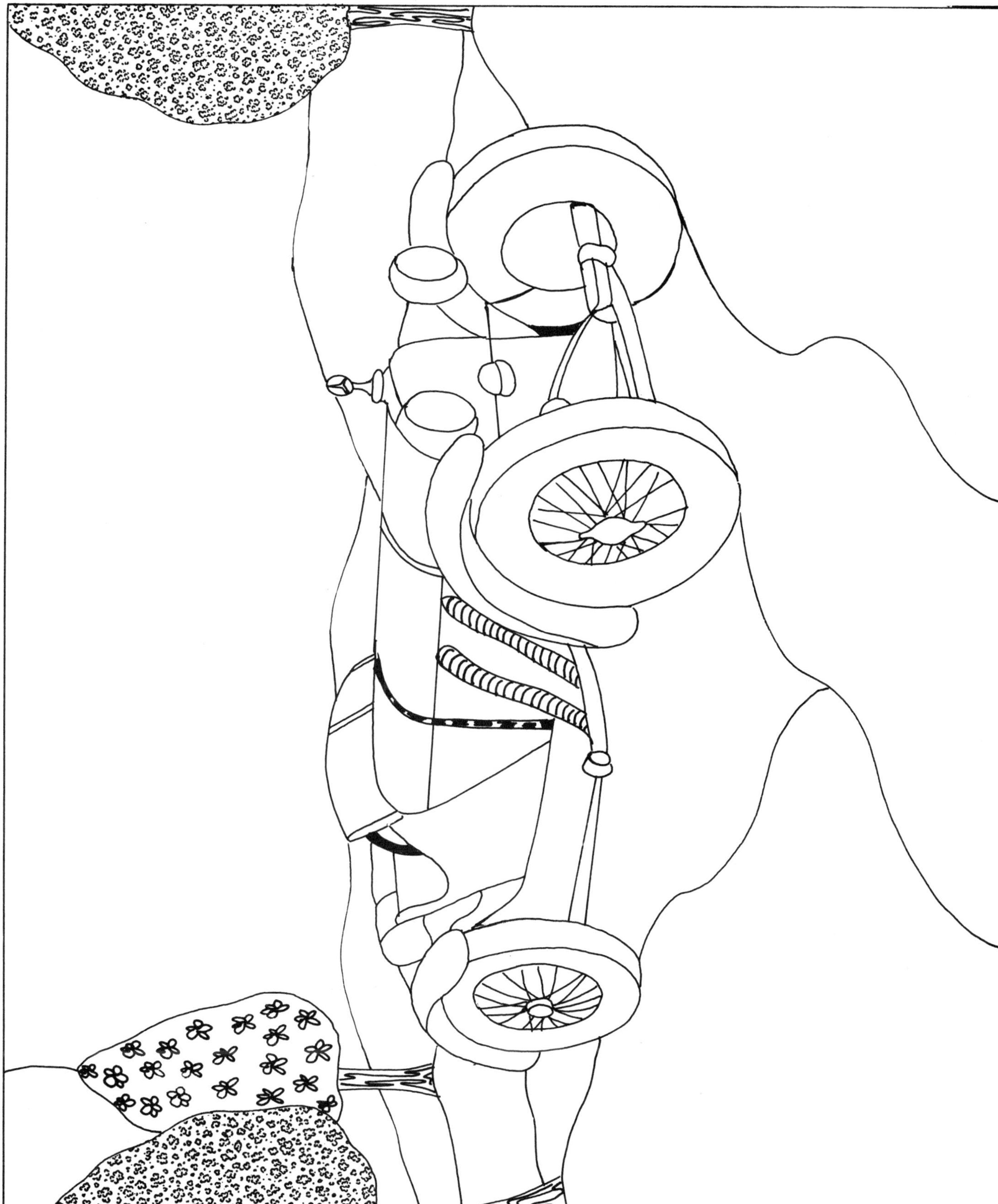

Night flight

Seagulls

Mandarin Duck

Canadian Geese

The Race

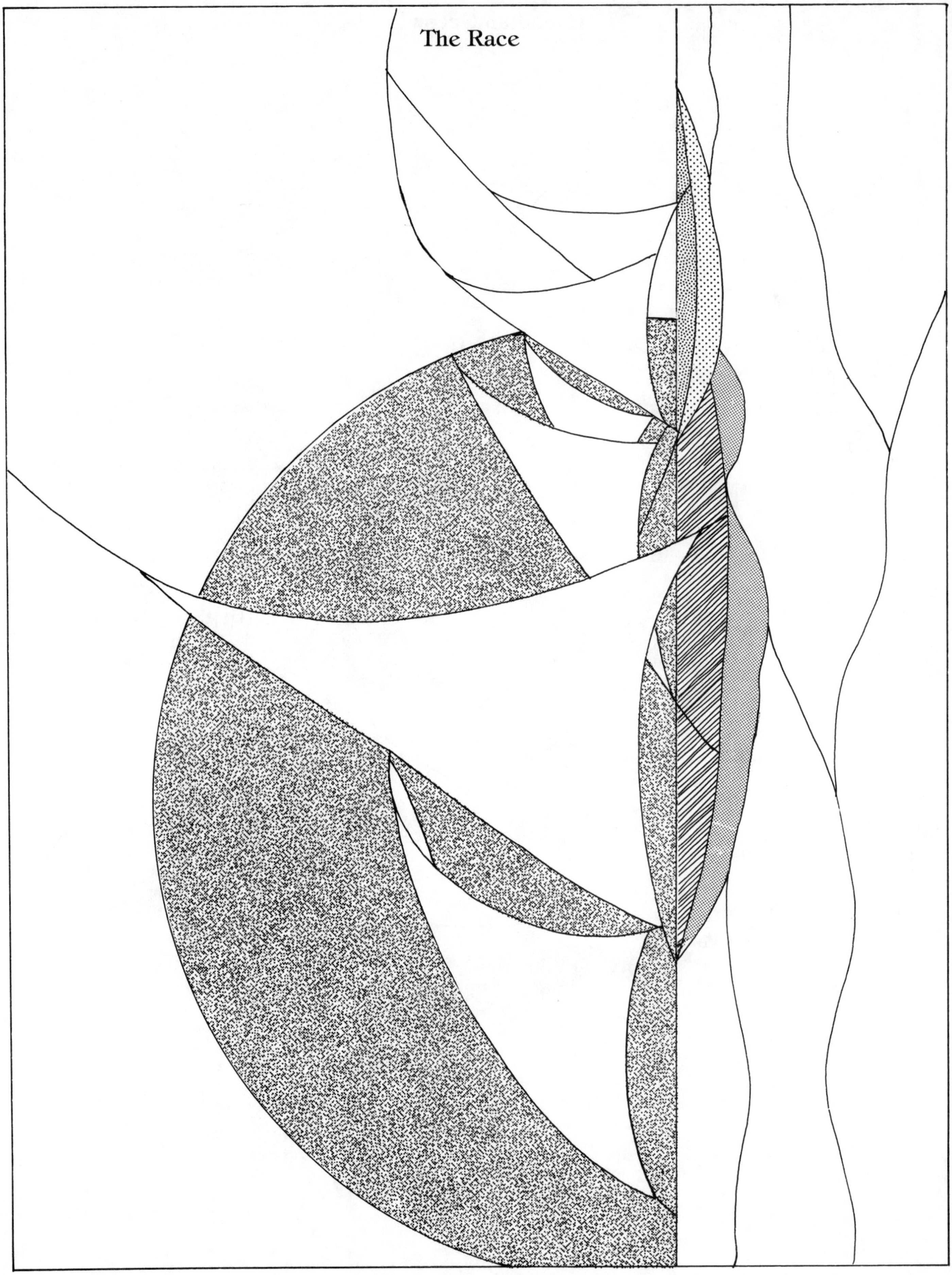

Sailing Ship

Elephant

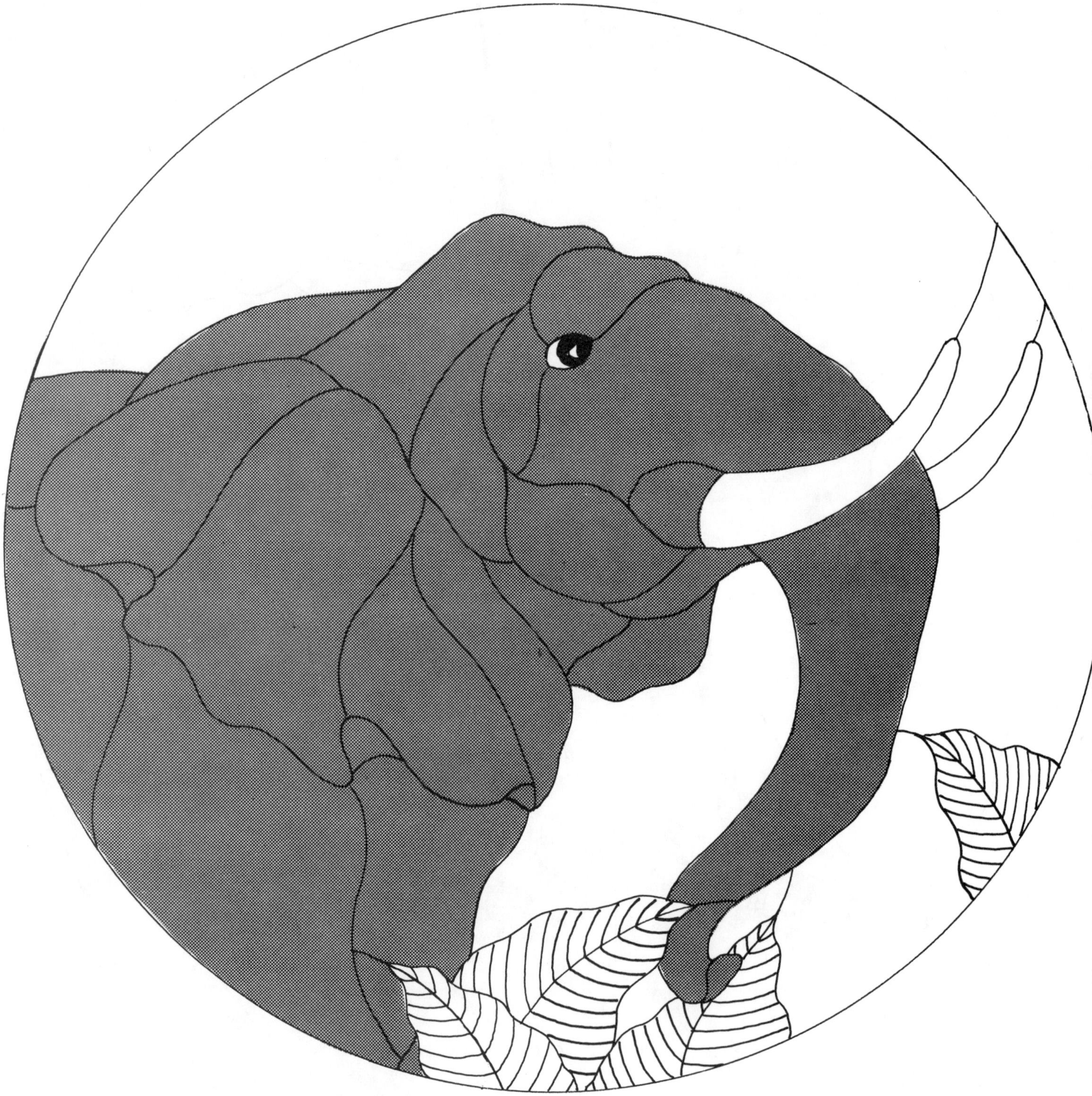

Moose

Eagle

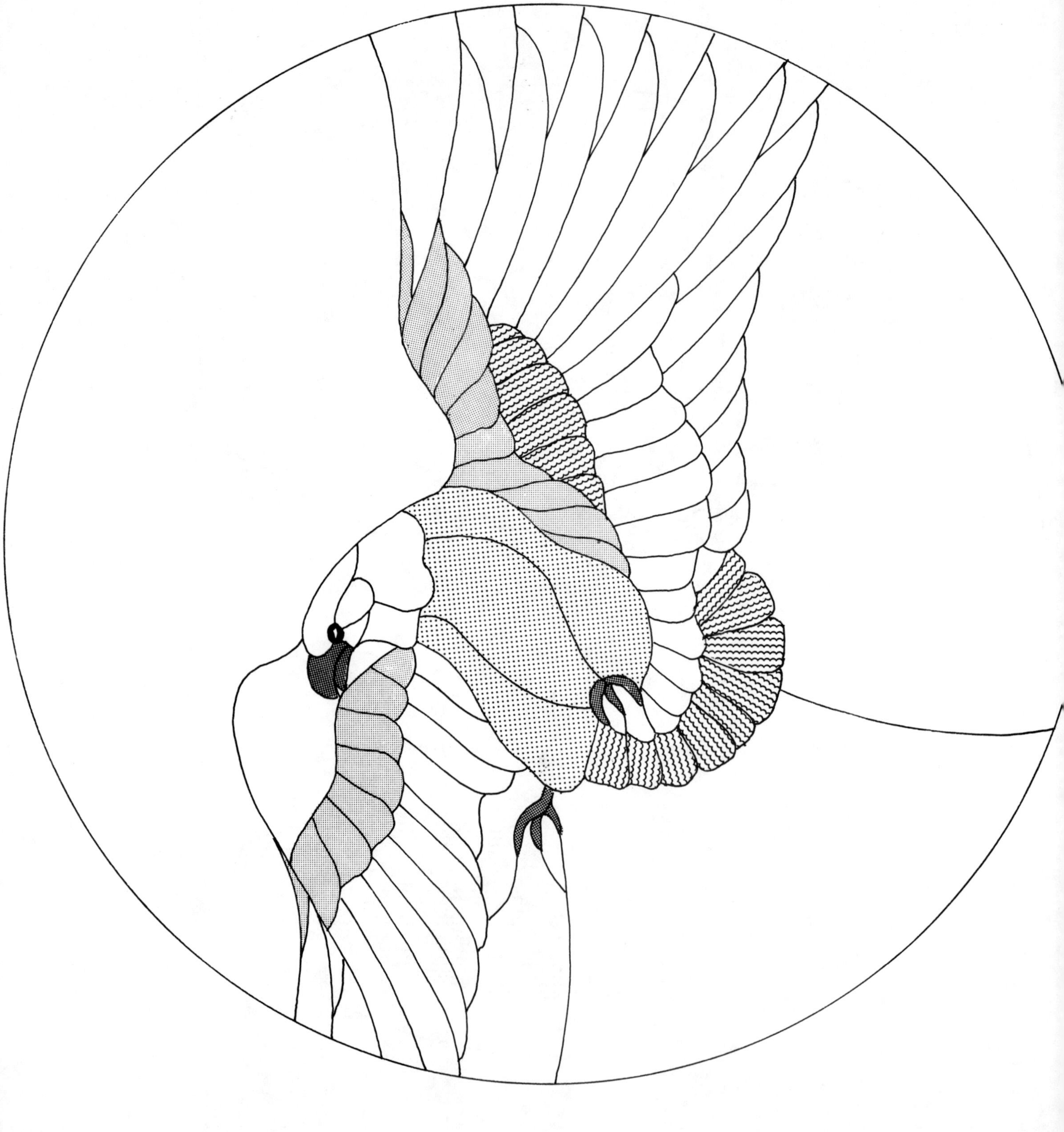

BATHS AND SPAS are ideal areas for designs of plants and animals associated with water. Lotus blossoms, swans, sea shells and seascapes are wonderful ideas. A section of a design borrowed from existing wallpaper, enlarged to fit the window requirements is another thought.

Baths and dressing rooms are wonderful places to add additional mirrors with stained glass overlays.

For a small bath, consider enlarging the room with a greenhouse window. These are very easy to install. Double glazing with a stained glass window will create a magnificent addition to the bath. If you do not plan to use the shelves for plants, there are no design restrictions for the window. However, if you retain the shelves and use the window as a greenhouse, keep the design simple. Use a maximum amount of clear glass (keeping privacy in mind) for the background so that the plants will have enough sunlight for growth. Warm pinks, peaches and ambers with a minimum of green seem to work best for plant growth.

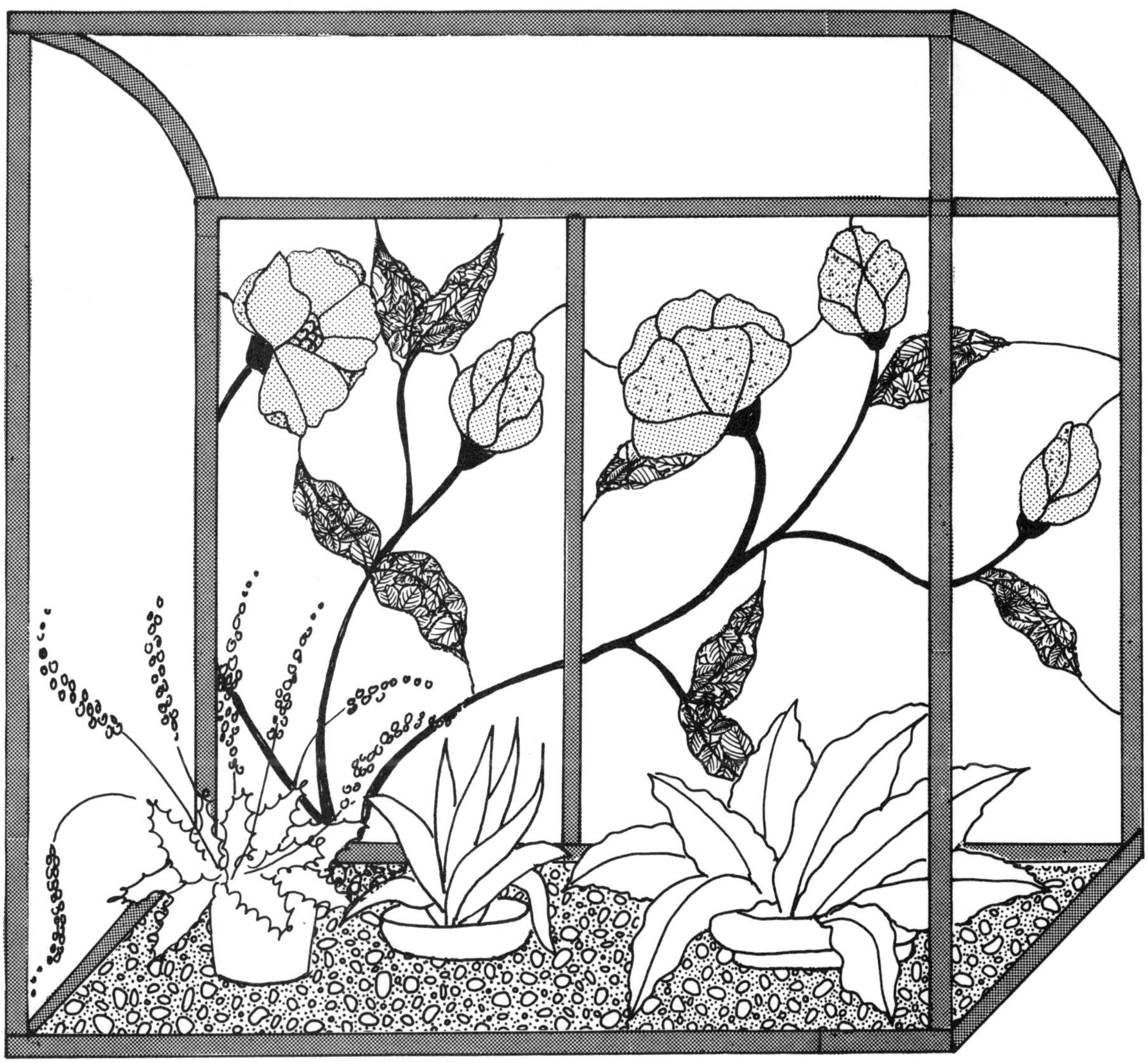

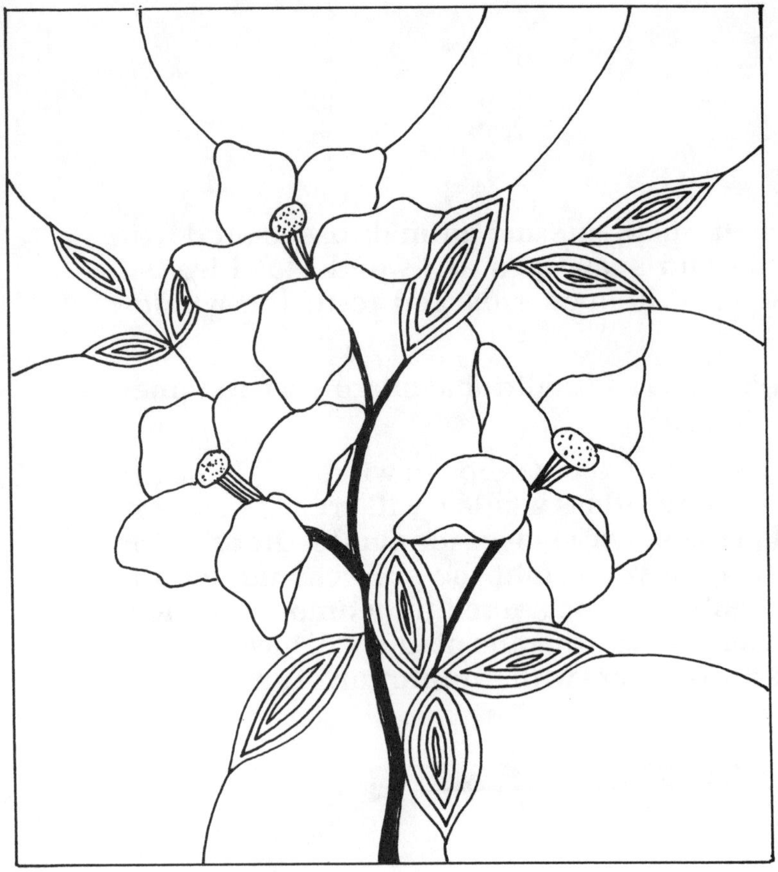

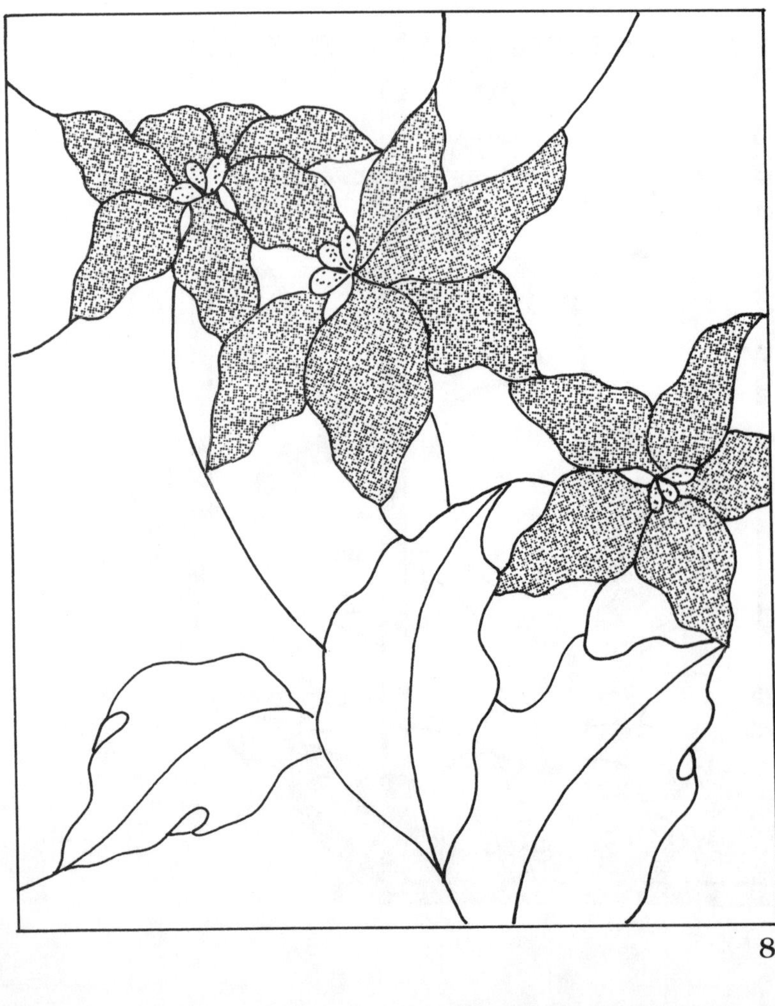

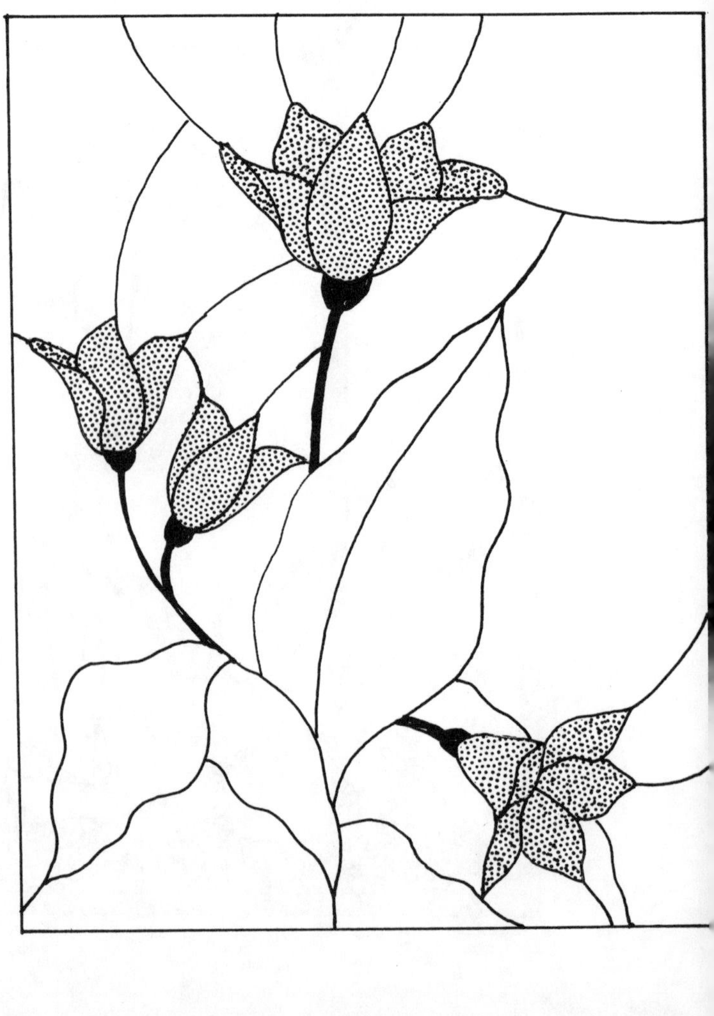

16th Century Court Fan

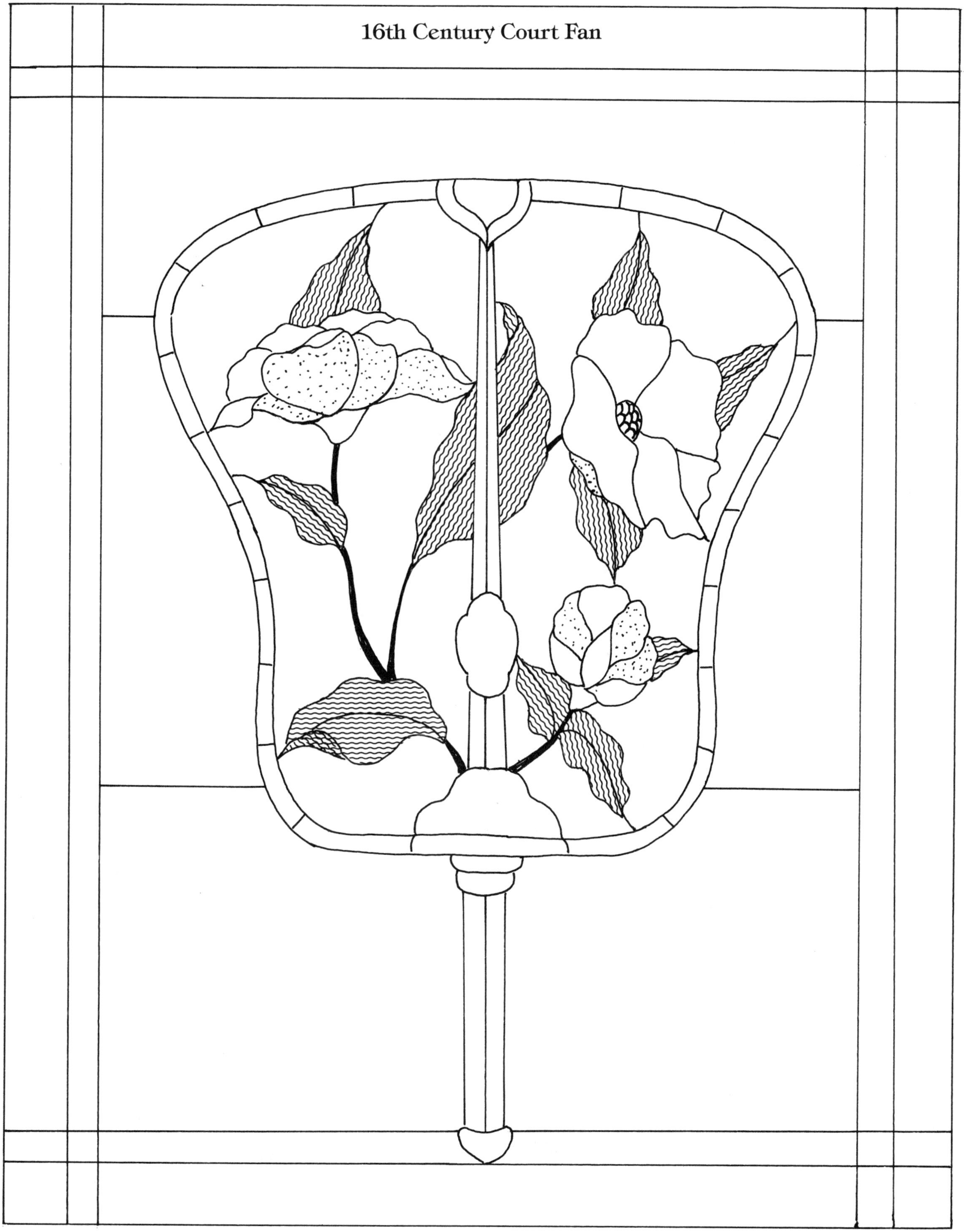

The Peacock

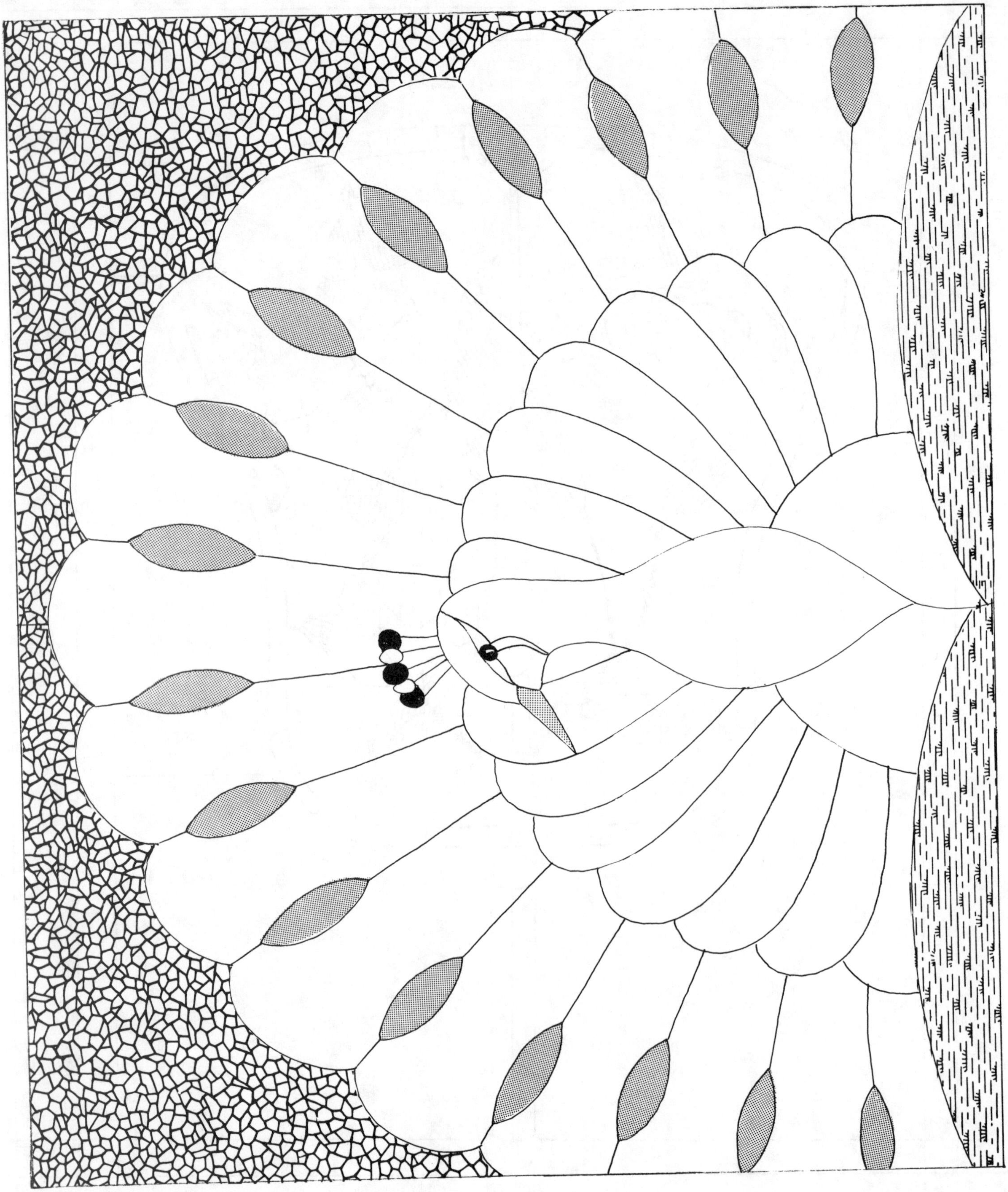

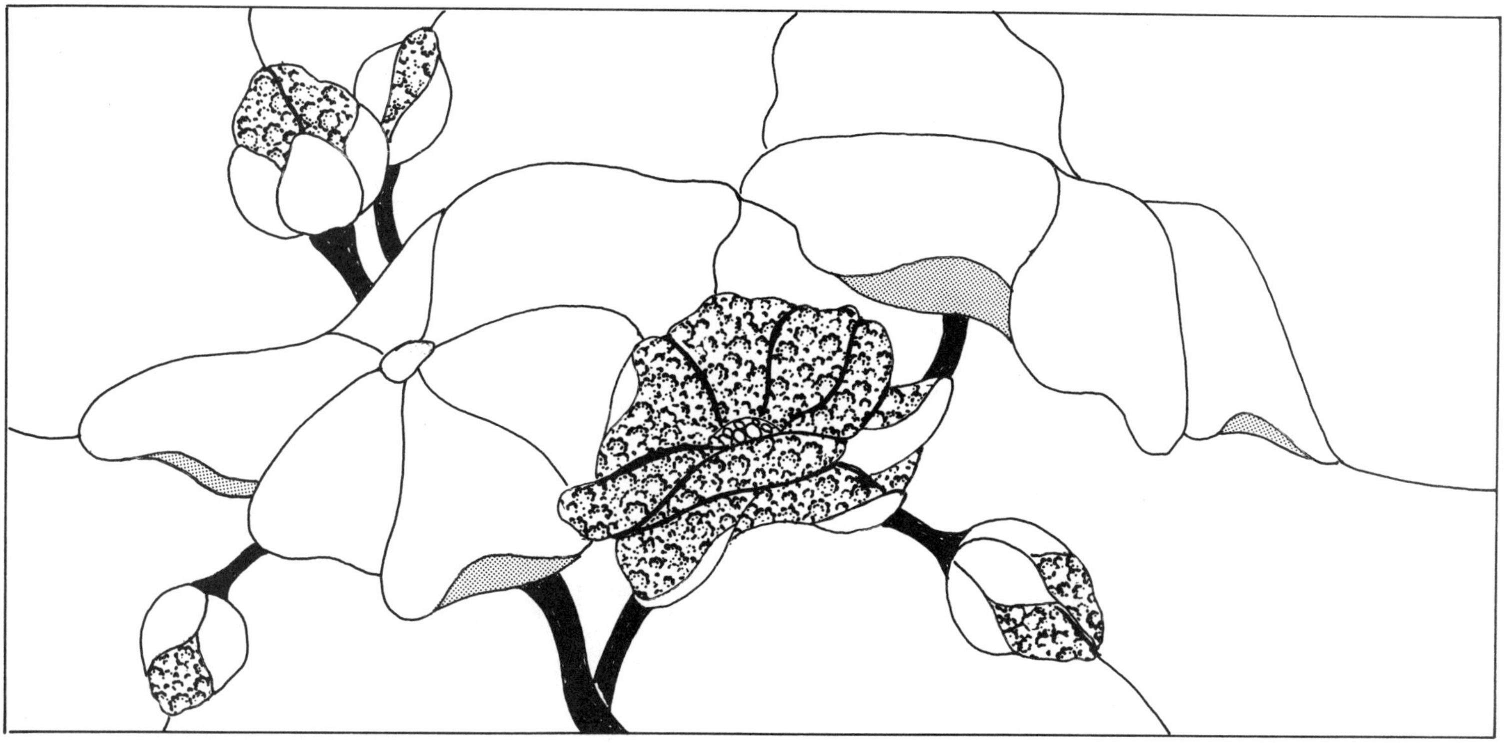

Lotus

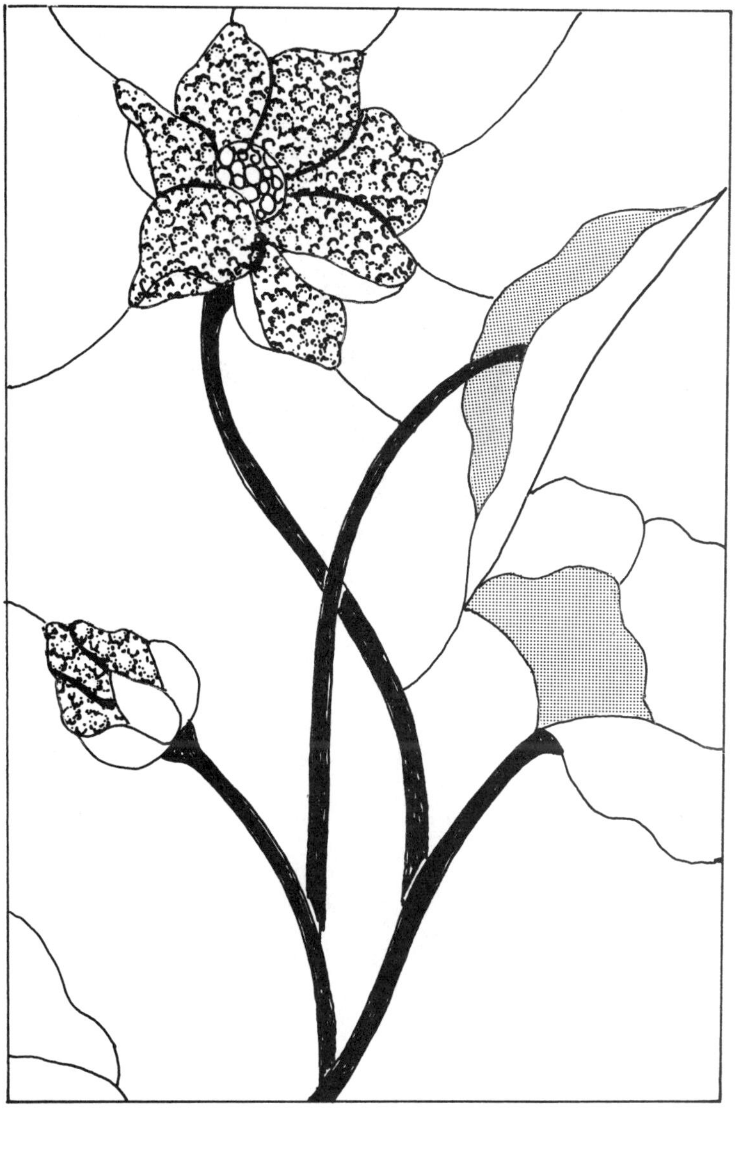

Mother Love

Tranquility

Morning Glories

Lilies

Mirror overlay

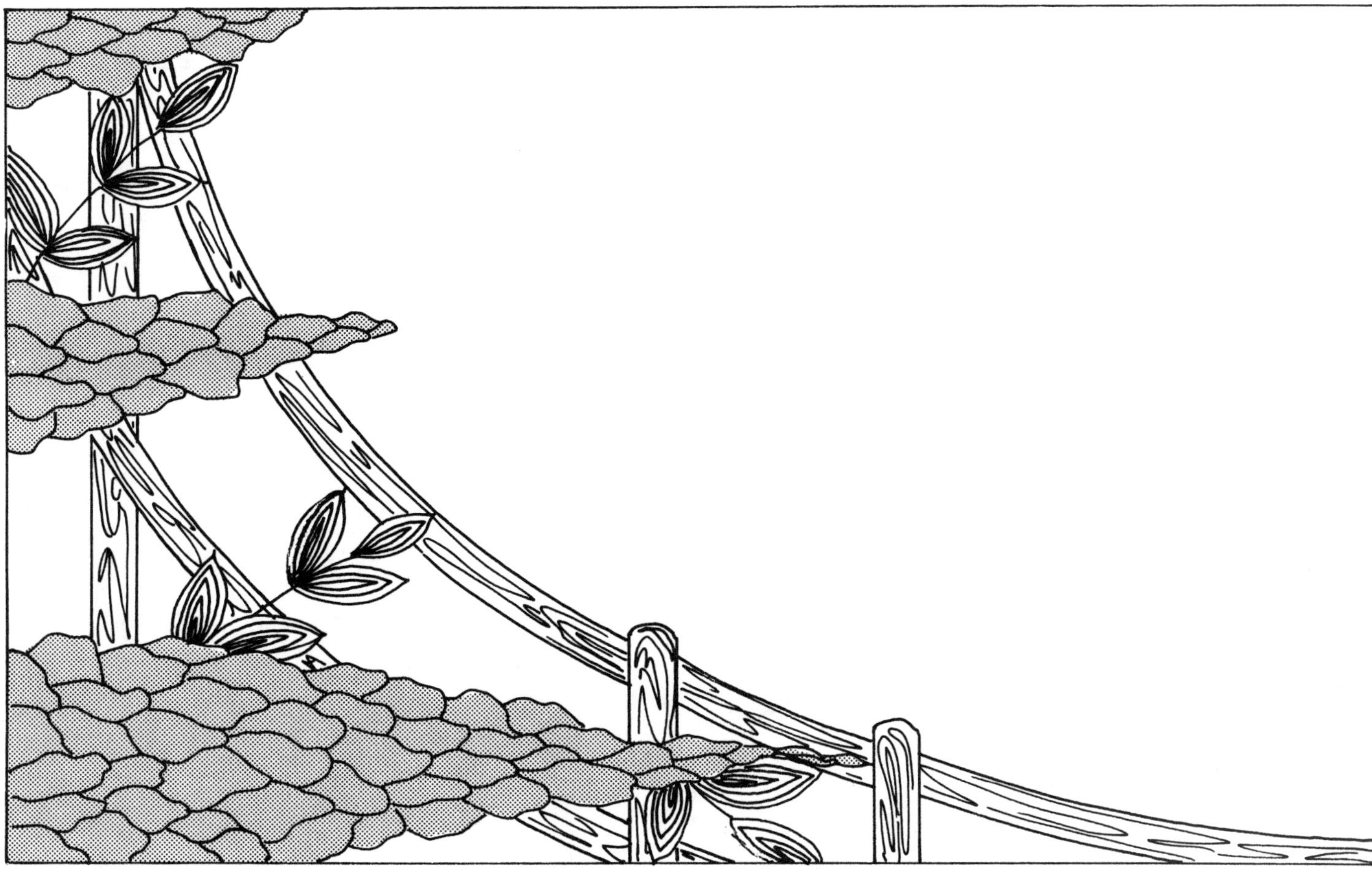

Mirror overlay

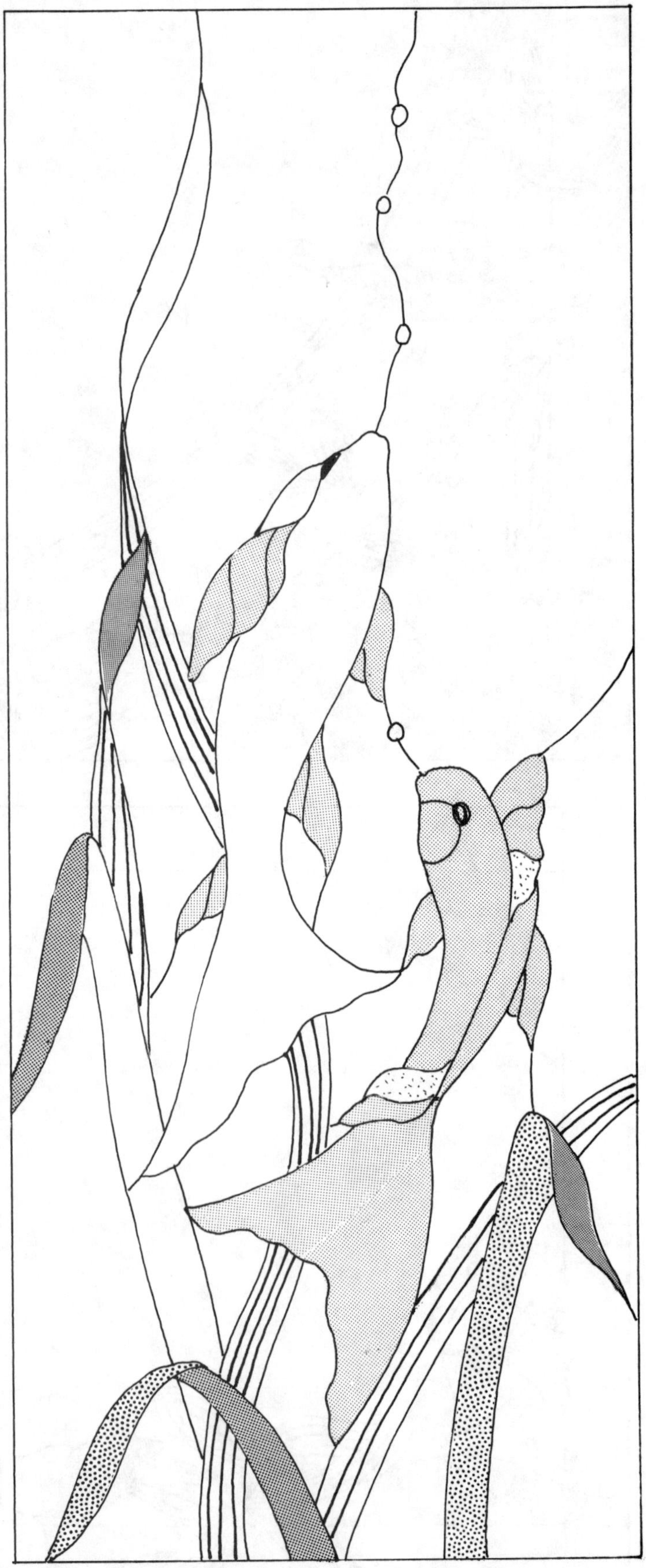

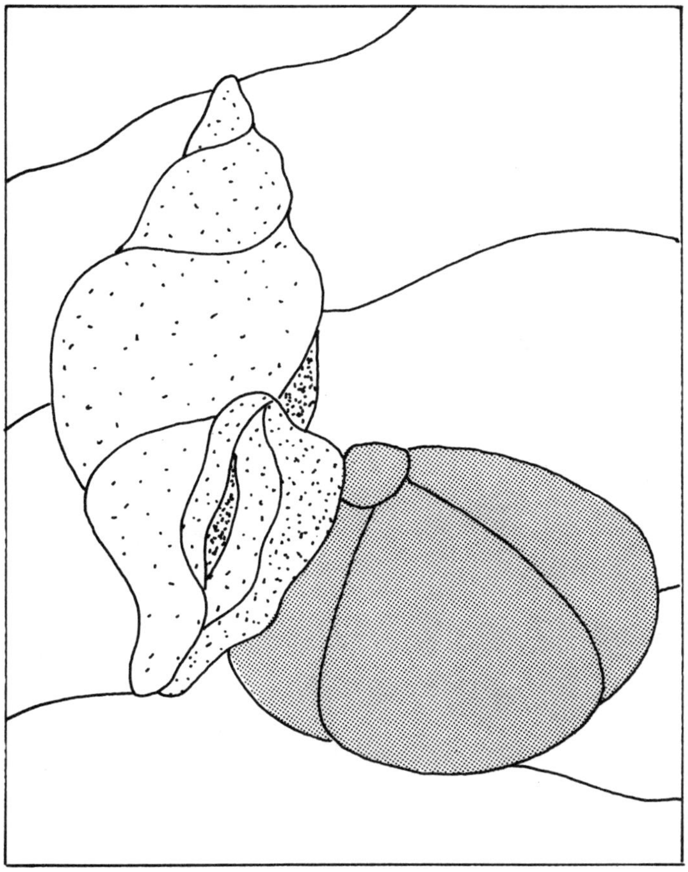

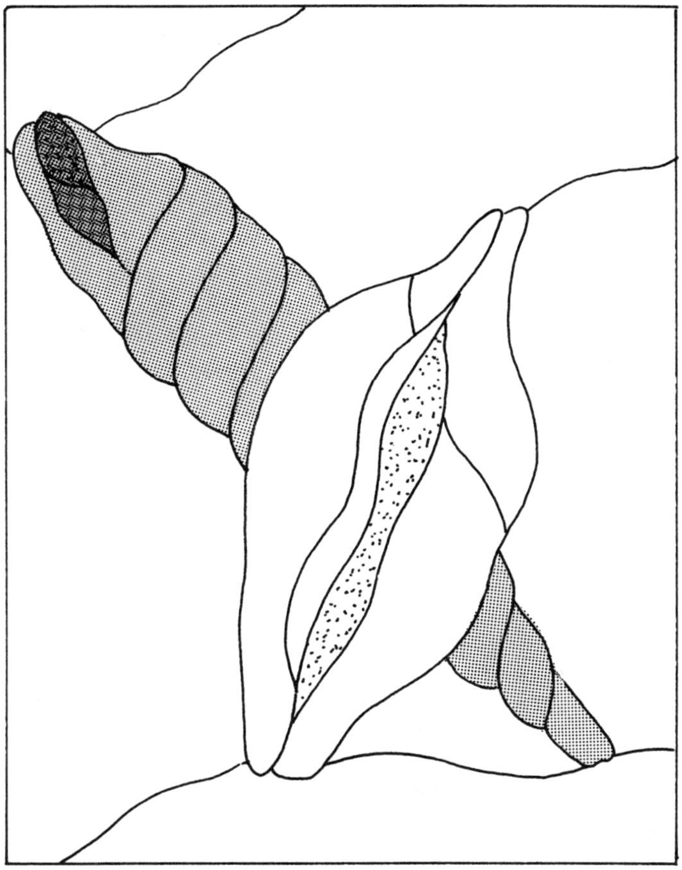

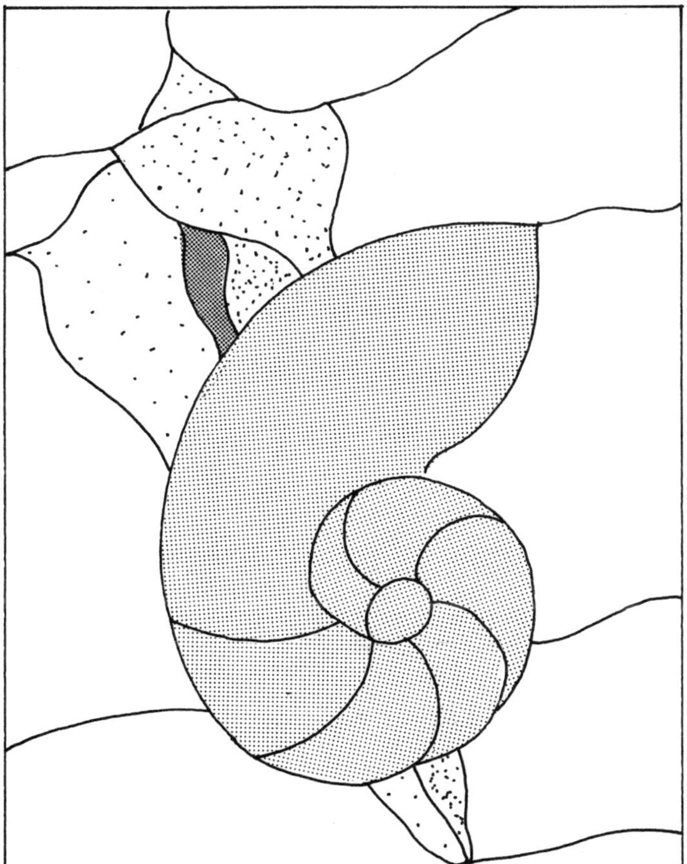

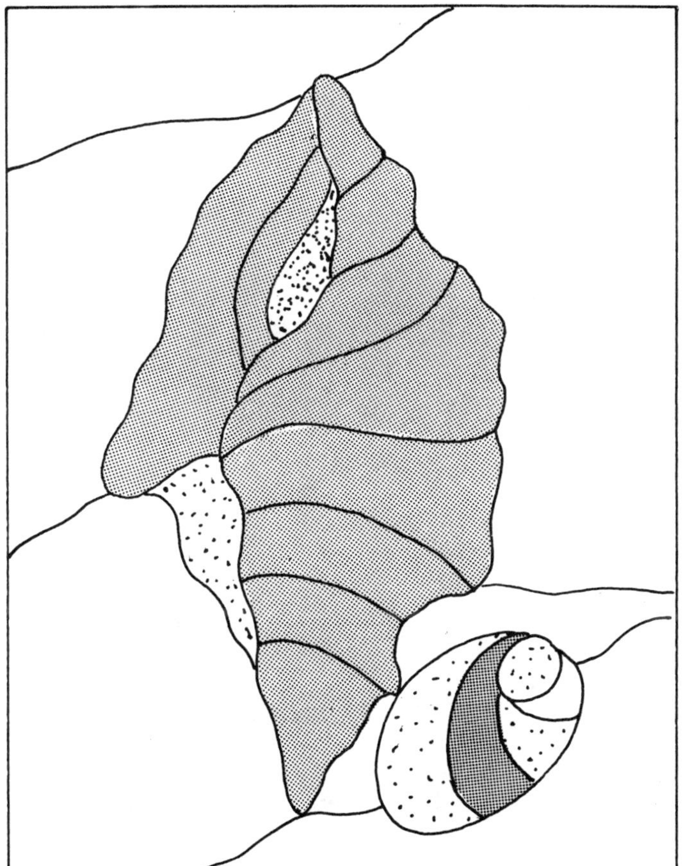

Flowers and Frog

Wisteria

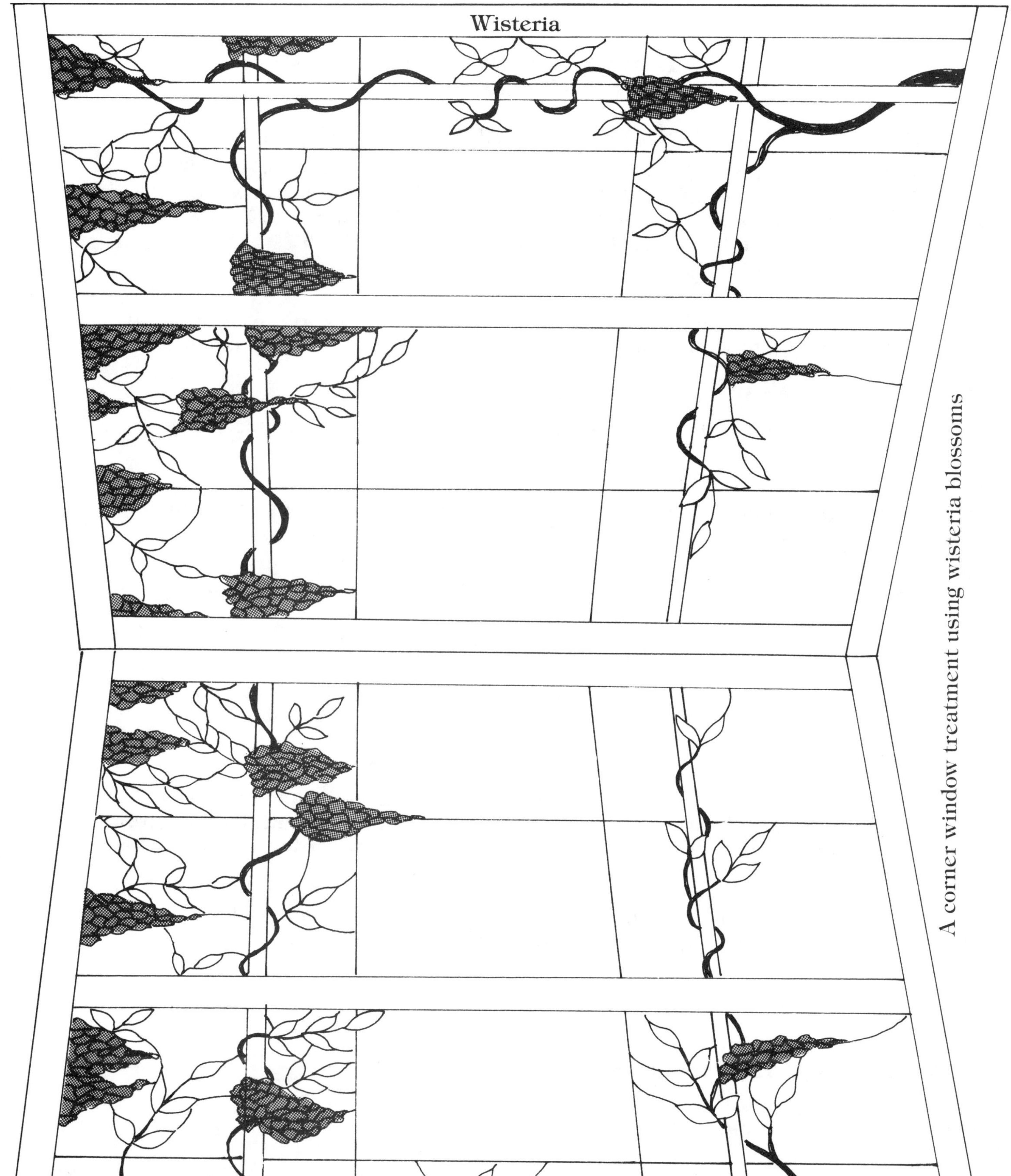

A corner window treatment using wisteria blossoms

ENLARGING THE HOUSE TOURS PATTERNS TO FIT YOUR SPECIFIC NEED

HT-1, HT-2, HT-3 & HT-4—SPRING, SUMMER AUTUMN & WINTER 11¾″x47¾″

Example: To increase to 18″x54″—custom bevels are omitted and stock bevel border is added. Increase width of pattern ⅜″ and length ½″. The pattern will now accept 2-2″x6″ bevels both top and bottom and 8-2″x6″ bevels on each side. Use 4-2″x2″ bevels for the corners. Add a 1″ outer border. Final size is now 18″x54″.

HT-5—WATER IRIS 11½″x37″

This would make an ideal sidelight alongside a door. A graphic shop can enlarge this pattern to fit an 18⅝″x60″ opening, a 22½″ x 72″ opening, or to your specific dimensions.

HT-6—THREE IRIS 25½″x34½″

Bevels control this pattern size but the outer border can be enlarged if necessary.

HT-7—WILD ROSES 22″ x 32½″

The right hand border and bottom border can be duplicated on the left side and top. This will give you a window size of 25¾″x36¼″.

HT-8—DOGWOOD & BUTTERFLIES 7½″x34½″ (each)

Enlarge by increasing border width or add an additional border.

HT-9—CAROUSEL HORSE 30″x30″ or 30″ Diameter

Outer border of desired width can be added. For a circular 30″ diameter window, omit four corners.

HT-10, HT-11 & HT-12 BIRD OF PARADISE "A", "B" & "C" 18″x28″

The addition of a bevel border will enhance and enlarge these windows. Adding 3-2″x6″ bevels to the top and bottom, 6-2″x6″ and 1-2″x4″ bevel to each side (the 4″ bevel in the center) and 4-2″x2″ bevels at the corners plus a 1″ outer border will give you a 24″x34″ window.

The three patterns work beautifully together to form a 54″x28″ horizontal window. Adding the above bevel arrangement enlarges this window to 60″x34″.

TO ORDER FULL SIZE PATTERNS

If patterns are not available at your local retailer, please send check or money order for $6.00 (includes postage and handling) to:

JUDY MILLER PUBLICATIONS,
2979 ALEXIS DRIVE,
PALO ALTO, CALIFORNIA 94304
(415) 948-7349